© **Copyright 2019 - All rights reserved.**

The content contained within this book may not be reproduced, duplicated or transmitted without direct written permission from the author or the publisher. Under no circumstances will any blame or legal responsibility be held against the publisher, or author, for any damages, reparation, or monetary loss due to the information contained within this book, either directly or indirectly.

Legal Notice:

This book is copyright protected. It is only for personal use. You cannot amend, distribute, sell, use, quote or paraphrase any part, or the content within this book, with   ͨ ᵗʰᵉ ᵃᵘᵗʰᵒʳ or publisher.

Disclaimer Notice:

Please note the information cᴄ   ɪs for educational and entertainment purposes only. All eɪɪɔɪɪ ɪɪᴀᴖ been executed to present accurate, up to date, reliable, complete information. No warranties of any kind are declared or implied. Readers acknowledge that the author is not engaging in the rendering of legal, financial, medical or professional advice. The content within this book has been derived from various sources. Please consult a licensed professional before attempting any techniques outlined in this book. By reading this document, the reader agrees that under no circumstances is the author responsible for any losses, direct or indirect, that are incurred as a result of the use of information contained within this document, including, but not limited to, errors, omissions, or inaccuracies.

Introduction *11*

Chapter 1: History, Benefits, and Tenets of Agile Project Management *18*

History of Agile 19

Advantages of Agile Project Management 22

Better Quality 23

Better Customer Satisfaction 24

Better Transparency 25

Better Control 25

Better Predictability 26

Better Risk Management 28

Better ROI 29

Better Metrics 29

Better Collaboration 30

Better Work-Life Balance 31

The Main Principles of Agile Project Management 33

The Agile Manifesto 34

The 12 Agile Principles 35

Chapter 2: Understand the Principles of Agile Project Management *38*

Customer Satisfaction 39

Making and Managing Changes 42

Continuous Customer Input 44

Daily Meetings 45

Sustainable Development 47

Continuous Improvement 49

Simplicity as a Vital Element 50

Self-Organizing Team 51

Inspect and Adapt..53

Chapter 3: Implementing Agile: How to Apply the APM Method Effectively *55*

Define Your Vision 58

Life Cycle Selection... 61

The Predictive Life Cycle 61

The Iterative Life Cycle.................................. 62

The Incremental Life Cycle 63

The Adaptive Life Cycle................................. 64

Creating an Agile Environment...........................65

Setting the Example 66

Being a Leader 67

Adopting Critical Thinking............................67

Stimulating Collaboration 68

Plan "Just Enough".. 68

Encouraging Face-to-Face Conversations 69

Creating the Physical Agile Environment 70

Delivery in an Agile Environment 70

Organizational Consideration for Project Agility........... 72

Call to Action 73

Chapter 4: Tools and Methodologies for Quality Control in APM...*75*

Clarizen 75

Trello...76

GitScrum 76

Jira 77

Taiga 77

Nostromo 78

Hansoft 78

Blossom 79

Ravetree 79

Chapter 5: Managing Risk in Agile Project Management. 81

Classify 82

The Descriptive Method 83

The PESTLE Method 85

Quantify 85

Plan 87

Act 89

Repeat 90

Chapter 6: Scaling Agile Projects 92

The Scaling Challenge 94

Scaling at Team Level 95

Scaling at Project Level 96

Scaling at Company Level 97

Agile Scaling Models 99

SAFe (Scaled Agile Framework) 99

DaD (Disciplined Agile Delivery) 100

LeSS (Large Scale Scrum) 101

Nexus 102

LeadingAgile 102

The Scrum of Scrums 103

Building Agile Teams 103

Be Patient...104

Adapt to Change 105

Be Results-Driven..106

Don't Miss the Bigger Picture.........................107

Be Transparent 107

Ask for Feedback 108

Give Feedback on a Regular Basis....................109

Show Trust — Always 109

Have Clearly Defined Roles.............................110

Encourage Team Members to Help Each Other 111

Make Your Office a Comfortable Place 111

Maintain the Stability of the Team..................112

Listen 113

Believe..114

Scaling Up: Agile Practices 115

Start with the MVP — The Minimum Viable Product 116

Use One Product Backlog 117

Make Sure Everyone Understands the Process 117

Emphasis on Collaboration 118

Trainings, Courses, and Certifications 119

Improve Your Development Infrastructure.....119

Stick to Smaller Teams....................................120

Coordinate the Production Time with the Iteration Length 121

Make Sure Someone Is the Product Owner 121

Synchronize the Iterations Across Teams........122

Use the Right Tools 122

Face to Face Meetings and Team Buildings 123

Scaling Out: Disturbed Projects 124

Conclusion.. 130

Introduction 140

Chapter 1: Agile Scrum Methodology..................... 149

What is Scrum Project Management? 150

What is Scrum in Relation to Agile Project Management?
.. 152

What is the Scrum Methodology Compared to Other Agile
Approaches?.. 154

Scrum vs. Kanban 155

Scrum vs. Extreme Programming (XP) 158

Scrum vs. Crystal.. 159

Scrum vs. Feature-Driven Development (FDD)............... 160

Scrum vs. Dynamic System Development Method (DSDM)
.... 160

The Scrum Team.. 162

The Scrum Master.. 163

The Product Owner 163

The Team Members (Development Team)...................... 164

Scrum Events (Ceremonies) 165

Sprint Planning 165

Daily Scrum... 166

Sprint Review 167

Sprint Retrospective 167

Scrum Artifacts 168

Product Vision 168

Sprint Goal...168

Product and Sprint Backlog 169

Burn-Down Chart 169

Increment 169

Scrum Rules 170

Practicing Scrum 171

Chapter 2: Lean and Kanban Software Development *175*

Main Principles of Lean Methodology...........177

Value 178

Value Stream 178

Flow 178

Pull..179

Perfection..179

Eliminating Waste 180

Amplifying Learning....................................183

Deciding as Late as Possible 186

Delivering as Fast as Possible 186

Empowering the Team.................................187

Building Integrity In 189

Seeing the Whole 191

How is Kanban Different from Scrum?.........192

Benefits of Kanban 195

Kanban and VersionOne...............................197

Chapter 3: Extreme Programming (XP) *200*

Planning Game 206

Small Releases 207

Customer Acceptance Tests 209

Simple Design 210

Pair Programming 211

Test-Driven Development 212

Refactoring 214

Continuous Integration 215

Collective Code Ownership 216

Coding Standards 218

Metaphor 220

Sustainable Pace 221

Chapter 4: The Crystal Method **224**

What Is the Crystal Method? 225

How Does Crystal Operate 227

Crystal Method Characteristics 233

Properties of the Crystal Method 235

Why Is the Crystal Method Useful? 236

Chapter 5: Feature-Driven Development (FDD) **238**

Domain Object Modeling 241

Developing by Feature 242

Component/Class Ownership 242

Feature Teams 243

Inspections 243

Configuration Management 243

Regular Builds 244

Visibility of Progress and Results 244

Chapter 6: Dynamic System Development Method (DSDM) *247*

Feasibility and business study 255

Functional model/prototype iteration 256

Design and build iteration 257

Implementation.. 258

Conclusion *260*

Agile Project Management

The Definitive Beginner's Guide to Learning Agile Project Management and Understanding Methodologies for Quality Control

By: Sam Ryan

Introduction

Project management is a real enigma to anyone working outside of this fascinating field.

On the one hand, project managers seem to be always on the run, quickly grabbing lunch before they jump into yet another report, yet another meeting, yet another book on how to scale agile at enterprise level.

Actually, trying to understand what all these armies of project managers do can be a baffling task. Some will simply dismiss the project management job and say they are mere email pushers. Others will glorify it and place project managers on a pedestal of golden spreadsheets and bossy tones of conversation.

As frequently happens, the truth is somewhere in between. Project managers are not email pushers, and this is something everyone in every single organization on Earth should understand.

They are the backbone of every business. They are those who move teams and make things happen. They are the ones who deliver products and implement feedback. They are the ones without whom companies can barely function.

At the same time, a project manager is not more important than, let's say, a developer or a QA (Quality Assurance) engineer. And a developer or a QA engineer are not more important than a project manager (PM). They are all part of the same team, aiming for the same goal: to make their customers happy.

Even when you go beyond all these misconceptions and stereotypes, project management can still seem enigmatic. It can still feel that PMs are people hiding behind endless spreadsheet colors and Jira tickets. It can still feel that project management is a job that can be done by just about anyone — after all, you simply run around and have conversations with clients, teams, and higher management, right?

The absolute truth is that project managers have to fit many gloves. They have to be excellent organizers and brilliant strategists, they have to be empathetic psychologists and business-minded leaders, and, ultimately, they have to maintain their own sanity even with all the constant running around.

Agile project management was born because there was a need for it — and this is something we will reiterate over the entire course of the book. Agile project management was born because project managers needed their sanity back and organizations needed their profitability back. As you will see throughout this book, agile came to solve a series of issues waterfall had been unable to address.

You should not take this the wrong way. Agile project management can be a saving grace in a large variety of situations — but it is not a magic code you apply to your work and automatically see everything improved. Agile takes hard work, and implementing it in a company that has been far too entrenched in traditional methods can be a real challenge.

Our purpose with this book is not to sell agile methods to you. It is, beyond anything, to help you see the real value in agile project management and everything it entails. Our book is meant to teach you the basics of agile so that you can incorporate it in your project management approach, in your team, and, finally, in your company.

We genuinely hope the book ahead will open new doors for you. If you are a waterfall project manager, you might find that this book will show you how agile can bring your skills up a notch and help you achieve more.

We also hope that this book will help you understand where waterfall is failing you, and why agile has grown to be so madly popular across organizations that are so diverse in nature and scope that it is almost impossible to ignore the growing trend.

If you are just starting out in project management, we hope this book will give you the knowledge to adopt a clean, clear, and efficient agile approach right from the very beginning. Agile project management can be a really productive and profitable job — on average, an agile PM in the US makes about $90,000/ year, for example (Glassdoor, 2019).

We will start off with a chapter dedicated to the very basics behind agile project management. Together, we will explore where agile comes from (and why there was such a great need for it to *happen* sooner, rather than later).

We will also explore the main advantages of agile project management. As you will see, this approach can change the entire landscape of your business — and it can transform your very own mindset, too. Agile can have a powerful influence across all business verticals. But, at the end of the day, it will simply make things better for everyone: for you as a PM, for your team members, for your company, and for your customers, too.

The second part of our first chapter is dedicated entirely to helping you explore the main principles of agile project management. Naturally, we will start with the absolute fundamentals of agile: The Agile Manifesto and the 12 Agile Principles, as they were documented almost two decades ago.

Once the foundation of agile has been established, we will move to the second chapter of our book. This chapter will allow you to explore the main Principles of agile and how they translate into actual agile practices — such as the (in)famous Daily Scrum, a mystery all non-agile teams have been trying to elucidate ever since they saw their organization's developers stand up for two minutes every day.

There is no point in knowing everything about agile history and Principles if you cannot actually apply this knowledge to your own business, right? As such, we will dedicate the third chapter of this book to the specific steps you have to make when you implement agile project management.

We will start with a vision and end with the organizational considerations and the call to action you need to develop before you jump in headfirst and create your first agile task board.

The fourth chapter of this book is all about tools. While there is a myriad of agile project management tools available, only some are used widely, across companies in a variety of industries. We will explore what these tools are all about, how much they cost, and whether or not it is worth giving them a try.

All projects come with a series of risks — and agile projects make no exception. Contrary to popular belief, being agile doesn't mean "not planning," it just means that you have a different approach to planning and everything it entails.

Agile risk management has its own specificities and we plan to explore them throughout the fifth chapter of our book. Together, we will move through the five stages of risk management in agile, so that you can create a comprehensive risk management plan before you start your project.

Last, but not least, we will dedicate the sixth chapter of our book to help you understand one of the biggest challenges of agile project management: scaling up. As you will see, there is quite a lot to learn here, so we advise you to make sure you understand the basic concepts elaborated throughout the first five chapters before you jump into the last.

The techniques and the concepts explained in our sixth chapter are of a more advanced nature, true. However, we believe it is absolutely mandatory for you to understand these ideas and how to tackle them — precisely because the agile implementation is, in itself, a scale-up effort. Even when you don't have to do it across a large organization, agile deployment and implementation means you will have to scale up your agile know-how and transfer it to your team, as well as coordinate with other teams within the same business to ensure a smooth flow of information between each group.

We don't want to lie: agile project management can be a challenge. However, when done right, agile can reshape the way you see work, the way you do business, and the very way you choose to live your life.

In theory, agile sounds like a dream come true — who doesn't love a work environment that focuses on honesty and transparency, right?

In practice, though, agile implementation might give you a good bunch of headaches. We are here to help you make sure you know how to "treat" these headaches, and even how to avoid them altogether.

Hopefully, by the end of this book, you will gain a deeper understanding of everything agile is — and everything agile isn't at the same time, too.

Hopefully, you will learn that agile is not something you can simply take as such and apply to your business. You will learn that agile is something you have to *work* with and adapt to your own needs. You will learn that agile does come with plenty of challenges — but that the vast majority of them are not in any way insurmountable.

Hopefully, you will learn that agile project management is more than just a set of rules. You will learn that agile can be an entire mindset and a life philosophy, and that you can apply it to everything — from the way you do your grocery shopping to the way you deliver the next big software application for medical research.

Hopefully, you will learn that agile is not chaos. In fact, agile is anything *but* chaos, precisely because it embraces change at its real value and it uses it as an opportunity to *be better*.

Waterfall project managers might not fully understand that agile is not about *not planning*, but about adapting your plan according to the changing circumstances. Because yes, change will happen, one way or another. And when it happens, it is far better to be prepared to adapt and embrace it, rather than stick to a pre-established plan that took no variation into consideration.

Our purpose here is to teach you to become better. It just so happens that the main method we will present here is agile project management. Regardless of whether you work as a PM in a software development agency or in any other industry out there, you will definitely find the lessons of this book to be valuable.

If you picked up this book hoping that you would find an infallible recipe on how to make agile work for you, we are sorry to disappoint. Our book won't promise that. And our honest advice on this is that *no* other book, training, or course will ever make that promise and actually deliver.

The very definition of *agile* has nothing to do with prescriptive methods. Agile itself was born to *fight* prescriptions — so if you ever come across any kind of material that promises to give you the golden recipe to agile project management success, we suggest you ignore it.

What we do promise, however, is that this book will change the way you perceive your work, your job description, and your company. It will change the way you see the world of project management, the world of business, and, in the end, *the world itself.*

It sounds grander than life, but agile has changed everything by enabling software companies to deliver real products that help real people. And when agile project management adoption moved past the software industry, the entire world changed its ways.

There is real transformation awaiting you at the end of this book. We truly hope you will be able to make the most out of these lessons!

Chapter 1: History, Benefits, and Tenets of Agile Project Management

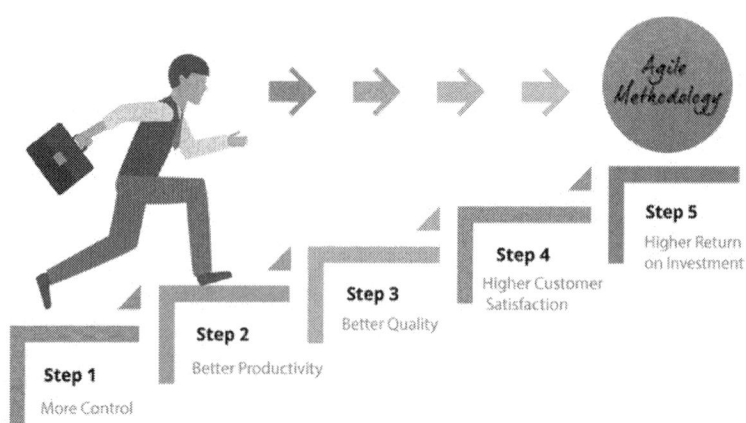

These days, it almost feels like *agile project management* is on everyone's lips. The term gets tossed around with so much ardor (among those who are *for* it and those who are *against* it) that we have almost lost its meaning altogether.

Agile is about *a lot* more than just hanging out in circles every day and playing little games when it comes to splitting larger tasks into smaller ones.

Agile was born with a reason — but even more than that, it was born in a time and *for* a time that needed a drastic change in approach.

Although agile methodologies (like Kanban, for example) have been practiced for many decades already, agile project management — in its raw, official format — was born a little under two decades ago.

In many ways, it feels like it's been ages already — and that is mostly due to the fact that agile has changed a lot. Its very foundation stays the same, as you will see throughout this chapter. But in terms of adaptability, agile has won the long-term game, and it is now one of the rulers in the world of project management.

How much of a ruler is it, really?

Studies show that about 30% of projects use a clear agile approach, while approximately 40% of them use a traditional/waterfall approach. The remaining 30% use a hybrid approach (Getapp, 2019) — which, contrary to what some may believe, doesn't show that waterfall has won the race, but that it needs agile adjustments to function in the modern world.

This chapter is meant to be an introduction into the world of agile project management — what it is, where it comes from, and why it is so popular. We invite you on a journey of learning and discovery, at the end of which you will discover which version of agile works best for you and for your organization.

History of Agile

As mentioned in the introduction of this chapter, agile project management was born out of sheer need. Together with the advent of modern technology and computers, a lot of things had to be done — and more importantly, they had to be done *fast*, accurately, and within the budget.

Agile methods started to take shape as early as the 1950s, but it wasn't until five decades later that the famous *Agile Manifesto* was laid on paper.

Sometime around the mid-'90s, most of what is now known as *agile project management* began to be contoured. At first, it was all dispersed — there was RAD (Rapid Application Development) in 1991, Scrum in 1995, and Feature-Driven Development two years later.

In 2001, it all came together, as if to offer clarity to a world that was getting ready to make a full jump into the one channel that rules them all today: the Internet. Sure, web connections had existed before, but it was at the turn of the millennia when things started to really take off.

On this background, seventeen brilliant software development minds got together in Snowbird, Utah. That meeting would go down in history, as this was when the actual concepts of agile project management took their official form.

We would be very curious to know exactly how that meeting went. We do know it didn't come out of the blue (you cannot simply expect the forward-thinkers in software development to just accidentally come together in Utah, can you?).

And we also know that the Snowbird meeting set the grounds for everything that followed in the world of agile project management. We would go as far as to assume that a lot of the amazing software tools we use today on a recurrent basis now would not have existed — they would have lingered somewhere, still in development.

Agile project management was born because there was a dire need to alter the project management landscape in software development. Back in the '90s, when the demand for software increased, traditional project management methods proved to be inefficient (at best) and downright disastrous (at worst).

In fact, it is frequently mentioned that most of the (major!) software development houses had (equally major!) lag when it came to their releases. On average, software development was lagging by approximately three years. For large projects, the lag extended further, up to 20 years.

It sounds preposterous, and it is.

But that's just what waterfall did to these projects. The very nature of software development is ever-changing, so it made no sense at all to stick to a management approach that focused on spreadsheets, more than the very core of what these projects did.

As you will see in the following section, agile project management has succeeded in removing those parts of waterfall that might have been bottlenecks in the development process.

Furthermore, as you will see throughout the entire book, the waterfall-agile hybrids have also ensured the streamlining of the waterfall processes, taking the best of the two worlds and providing organizations based on heavy documentation with an alternative to traditional project management.

Agile project management was born at the right time to become the backbone of the tech industry before it boomed completely.

From the Silicon Valley to China, and from Iceland to Australia, agile has become a household name in project management. Even more, it has expanded well beyond the borders of software programming and is now used in pretty much every industry you could imagine. Hospitals, schools, governmental and non-governmental institutions, marketing, translations — everyone can embrace agile, precisely because it is so flexible and adaptable to multiple situations.

History is rarely written when we expect it. We don't know if those seventeen minds actually knew the kind of impact their meeting would have — but we do know that software development and agile are so intertwined these days that it seems almost impossible to completely separate them.

Advantages of Agile Project Management

Enough with the praises, though!

If you are reading this book, it is quite likely that you have already heard about how great agile is and the "miracles" it can bring about.

What makes agile project management so good, specifically?

There is a long list of benefits that bring agile project management to everyone's attention — specifically those in software project management, but most definitely not exclusively so.

Better Quality

One of the main tenets of agile project management is that it promises a better product quality.

To get things straight, it's not that products developed via waterfall project management lack in quality. It's just that it is far more likely for an agile-developed product to be qualitative by the point of its full release on the market.

There is very strong logic behind this.

On the one hand, waterfall project management tends to be too strict within its own limits. This means that it is far more likely for mistakes to:

- Be noticed too far out in the process (and waterfall will not allow the team to reiterate the same feature/part of the project)
- Be noticed when the product is already on the market/in review by client

Agile project management is *very* focused on continuous improvement. As such, a product has a much higher chance of actually getting better throughout the development process.

Or, in other words, instead of sweeping all those small (or not so small) mistakes under the rug (as you would do in waterfall project management because you have to follow the *plan*), you will just *deal with them there and then*.

Sounds much more feasible, right?

Better Customer Satisfaction

Another reason that makes agile so advantageous is related to the fact that, by the end of the project, customers tend to be far more satisfied.

How so?

There are a few verticals to consider when it comes to customer satisfaction, and agile makes sure that all of them are properly met. For instance:

- Agile will allow you to change the requirements as per client feedback.
- Agile will force you to release bits of the project as you move along, ergo, it will allow your customer to provide you with input that is easier to implement (due to the small size of the actual bit they are testing).
- Agile will help you deliver a better final product (as shown in the previous section).

Given all these factors, it makes all the sense in the world that customers will be happier — throughout the project, as they will be able to request the modifications they need and, at the end of it, as they will receive a product that fits their requirements, purposes, and desires.

Do keep in mind that the same stands true in those cases when the "customer" is an internal stakeholder — such as, for example, when you are managing a software development project meant to be used internally.

Better Transparency

This agile benefit is very tightly connected to what has been mentioned already. When you can ensure better quality and better customer satisfaction, it all comes with greater transparency.

This transparency will manifest itself on all the verticals of project management. You will see better transparency in your team (and, as you will see later on in the book, agile has developed internal mechanisms and tools you can use to ensure this happens).

You will also see better transparency within the organization, regardless of whether or not the upper management uses the same project management approach as you do.

Finally, you will see better transparency between you and your customer (be it an internal one or an external one). When you constantly ask for feedback and continually improve the product to suit your customer's needs, you create a more genuine relationship with them. You start to truly communicate, rather than engaging in nothing more than ping-pong emails.

Better Control

For those of you used to the premises of waterfall or traditional project management, it might seem that agile is anything *but* control-focused.

In fact, it very much is.

Agile project management allows you to control your project at a granular level, precisely because it encourages (and downright forces) you to split your project into small, bite-sized bits and pieces.

Waterfall project management forces you to lay it all down on paper before everything begins. At the same time, though, it also forces you to stick to the plan even when things go south. And yes, they will eventually go south, one way or another: the client's requirements might change, you might realize something is taking longer than planned, your product might be bugged, or the costs might end up exceeding your expectations.

There are a million things that could go wrong, especially in software project management (where things tend to be more experimental than, let's say, oil drilling, for example).

When you can manage all these things that could go wrong *as* they happen, you gain control over the entire process. Even more, you can use your (bad) experience to improve the process, as well.

Don't take this the wrong way. *Control* is not one and the same as *micromanagement*. You don't have to constantly look over your team's shoulders and manage each tiny detail every step of the way. That would just ruin the bridge of transparency, honesty, and self-discipline you are trying to build between yourself and your team.

Better Predictability

Again, this might seem like it's the exact opposite of what agile project management is all about.

But when you take a closer look, you will realize that agile projects can be better predicted precisely because they are managed step by step.

Let's compare this with baking a cake.

When you buy the boxed mix, you can easily and accurately predict what you are going to get — a decent cake batter you can then personalize according to your tastes. However, you don't know all the ingredients in that boxed mix — and, although the short-term result might be easy to foresee, it might be a bit more difficult to predict what will happen to your body if you continue eating boxed cake every week, for decades on end.

That would be the waterfall project management approach. You are using a mould and hoping that everything in your project will fit that ideal, very predictable format. However, long-term, you have no idea if your project plan won't go against you.

When you bake a cake from scratch and you know where each ingredient comes from, how many calories it has, and how many nutrients it provides your body with, you can predict its effects on your body if you eat the same type of cake for an extended period of time.

Plus, as long as you accurately measure each ingredient's quantity, you will be able to accurately predict how your cake is going to look and taste like. It might take a bit of practice until you learn how to do this correctly but, once you learn its tricks, the cake made from scratch will be more predictable in every single way!

That would be the agile project management approach. It might seem totally unpredictable at first, but the results will become more predictable once you have the right tools and the experience to accurately measure and approximate everything.

Better Risk Management

One of the major downfalls of waterfall project management is connected to the fact that it remains confined within its own tables and spreadsheets.

Waterfall project managers plan everything out at the beginning of the project. Agile project managers do the same. The main difference lies not in *whether* they plan, but in what happens when things don't go according to that plan.

As mentioned before, waterfall tends to sweep risks under the carpet — or, at least, estimate them poorly and through an idealistic point of view.

Agile, on the other hand, doesn't do that. It faces the problems head-on, tackles them, removes them from your path, and then allows you to draw honest conclusions.

As a result, your risk management will improve, as well. When you stop hiding your head in the sand, you can see things more clearly. As such, you can manage any potential risks with more accuracy, as well.

Better ROI

Put every advantage we have already discussed throughout this chapter one on top of the other and you will understand why agile project management tends to lead to better ROI.

Better products + happier clients + better risk management cannot go wrong.

It's a universal formula for success. The more you can manage your money and put out better products, the more likely it is that customers will:

- Come back to you
- Pay on time
- Evangelize and recommend you to other potential customers
- Leave great reviews for your company on various channels

Sounds like a dream?

We prefer to call it *agile*.

Better Metrics

This advantage circles back to the fact that agile won't allow you to just sweep problems under the carpet. It will make you have a face-to-face conversation with these issues, get to know them in-depth, and then tackle them from a stance where you actually know what to do.

Plus, agile project management is a team effort in every respect. From the moment you start splitting your project into smaller chunks, your team will be involved in the process. They will be able to give you real-life estimations on how long everything takes.

Finally, agile project management will allow you to track what is *genuinely* going on, rather than what you idealistically projected to happen.

All of these aspects will eventually lead to better, more accurate, more realistic, and more useful metrics when it comes to team performance, ROI, and time management.

Better Collaboration

If there is one thing absolutely everyone loves about agile project management (aside from the apparent chaos, which, by the way, can become addictive) is the fact that teams just tend to work better when they are managed under an agile method.

Agile project management fosters an environment that focuses on self-discipline, honesty, and taking responsibility. When you have these three ingredients, you create *true* team spirit — the kind where people naturally understand and empathize with each other, where they genuinely want to help each other, and where various types of frustrations and bad feelings don't even take root.

Agile is all about collaboration. The way you collaborate with your team, the way your team members will collaborate among themselves, the way your product manager will collaborate with the client, and the way you will collaborate with other stakeholders and upper management within your company — this will all change for the better.

This is not an empty promise. It lies at the very foundation of what agile is and what this approach aims for.

Better Work-Life Balance

We won't lie.

Not all people who work in agile project management have a great work-life balance.

But, then again, not all people who work in *anything* have a great work-life balance.

It is generally believed that those who work in agile project management (meaning the project managers *and* the teams) tend to have a better work-life balance because they learn how to efficiently manage their time. Therefore, they are much less likely to slack off and prolong their workdays into work nights and work weekends.

They are more likely to get their job done in the time it is supposed to be done — so that in their off hours, they can go back to their families, hobbies, and spare time.

Overall, this can lead to nothing but better, happier, more productive employees.

And we all know how happy *that* makes management, HR, and every single part of your organization, right?

You don't have to take our word for it when it comes to all these benefits. You just have to look at those companies that have embraced agile as part of their structures — they have plenty to say about it and how it has drastically changed their entire way of doing business.

These are just some of the advantages. You might experience all of them, a few of them, or more. In any case, you will definitely enjoy a noticeable, realistic improvement in the way your projects are managed!

The Main Principles of Agile Project Management

12 AGILE PRINCIPLES

01 Our highest priority is to satisfy the customer through early and continuous delivery of valuable software.	**02** Welcome changing requirements, even late in development. Agile processes harness change for the customer's competitive advantage.	**03** Deliver working software frequently, from a couple of weeks to a couple of months, with a preference to the shorter timescale.
04 Business people and developers must work together daily throughout the project.	**05** Build projects around motivated individuals. Give them the environment and support they need, and trust them to get the job done.	**06** Agile processes promote sustainable development. The sponsors, developers, and users should be able to maintain a constant pace indefinitely.
07 Working software is the primary measure of progress.	**08** The most efficient and effective method of conveying information to and within a development team is face-to-face conversation.	**09** Continuous attention to technical excellence and good design enhances agility.
10 Simplicity – the art of maximizing the amount of work not done – is essential.	**11** The best architectures, requirements, and designs emerge from self-organizing teams.	**12** At regular intervals, the team reflects on how to become more effective, then tunes and adjusts its behavior accordingly.

Agile project management did not come out of the blue.

Those seventeen people didn't just sit down over beer and chips and lay out whatever crossed their minds.

As we have already mentioned, each of them had already experimented with various forms of what they called "lightweight project management" (the term "agile" was more informal at that point, rather than official).

Some had experimented with Kanban, others had experimented with Scrum, and others had experimented with other types of agile project management over the course of the '90s.

When all of it came together, these bright minds came to a conclusion and created what is now known as the *Agile Manifesto.*

There are four main points to the Agile Manifesto — and all of them are mirrored in the *12 Agile Principles.* Together, the Manifesto and the Principles make up the "bible" of agile project management — the nearly sacred rules that lie at the foundation of everything agile under the sun.

We won't dwell too long on either the Manifesto or the Principles, as they will both be emphasized and explored in-depth throughout the following chapters of the book. We do want, however, to give you a full picture of what the fundamental "laws" of agile look like — so we will briefly move through the Agile Manifesto, and then through the 12 Agile Principles before we jump into the second chapter of our book.

The Agile Manifesto

As mentioned above, the Agile Manifesto is comprised out of four major rules. Together, they lay the groundwork upon which agile is built, starting with its foundation, the *principles.*

The four rules of the Agile Manifesto are simple in nature:

1. Interactions and individuals are always more important than tools and processes.
2. It is far more important to deliver working software than to document every single step of the way.
3. You should collaborate with your customer, rather than negotiate and re-negotiate your contract.

4. You should be ready to respond to change, rather than strictly adhere to a plan.

You can read the Agile Manifesto in its initial format, as well as its original signatories at https://agilemanifesto.org/

The 12 Agile Principles

If the Manifesto is the ground upon which agile is built, then the *Principles* are its foundation. Brick by brick, these principles are applied in every single type of agile project management. From Scrum to Kanban and from XP to Lean Software Development, these principles are valid across the entire agile spectrum and its large number of variations (including many of the so-called *hybrid* approaches).

The 12 Agile Principles state as follows:

1. You should deliver working software in a continuous way and always on time (if not earlier). This will improve overall customer satisfaction.
2. Changes are welcome, even when they come late in the development process.
3. You should always make sure to deliver working software in batches, every few weeks. Generally, you shouldn't allow *months* to pass between your deliveries.
4. Business people and developers within the organization should collaborate and cooperate on a daily basis.

5. The individual team members should be motivated and the projects should be built around them, with trust as a core concept of the entire management approach.

6. Avoid long-distance communication as much as possible. As you will see, this rule has changed and has been adapted to modern times and remote teams, but it is still highly encouraged.

7. Measure your progress by how much working software you deliver, rather than any other metric.

8. Make sure your plan is sustainable and that your entire team and organization will be able to maintain a constant pace.

9. You should make sure you always pay attention to technical excellence and design that is genuinely good and easy to understand by the end user.

10. Keep things as simple as you can and focus on the work *not* done. In other words, minimize the work you have to do by eliminating those tasks that are not absolutely necessary.

11. Self-organized teams will always lead to the best results: better architectures, better designs, better adherence to the customer requirements.

12. Your team should always reflect on how they are doing (and what they are not doing well), so that they can adjust accordingly.

Same as with the Manifesto, you can find the 12 Agile Principles in their original format at: https://agilemanifesto.org/principles.html

As mentioned before, we will not dwell on the Principles (or the Manifesto) at the moment. They are the beating heart at the center of agile and they are certainly important, yes. But it is also important to know how to interpret them correctly and how to apply them to real-life situations.

As such, the second chapter of our book will be dedicated to exploring the Principles and understanding them. Most importantly, though, it will focus on helping you learn how to apply these tactics into actual practice.

In other words, we're getting to the *really* interesting part!

Chapter 2: Understand the Principles of Agile Project Management

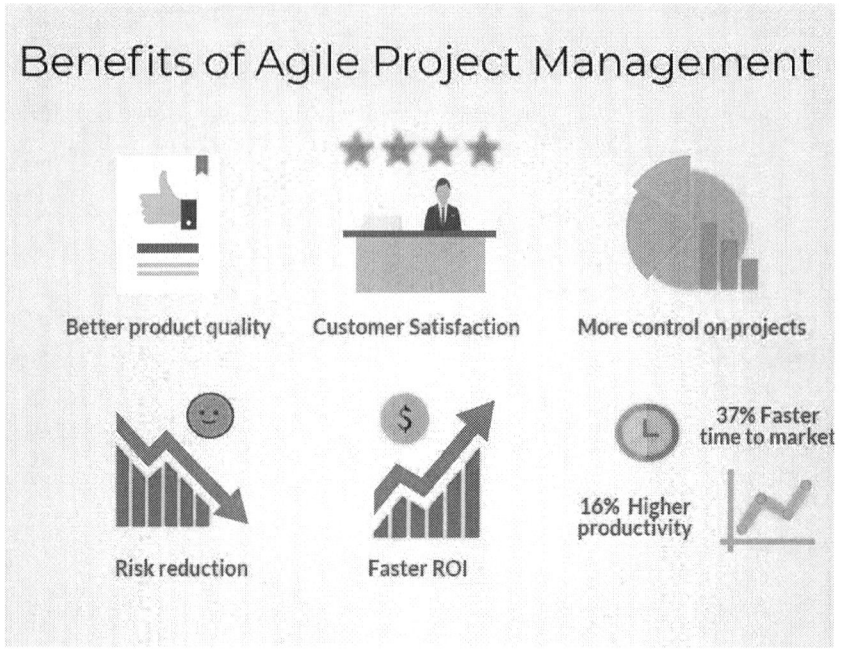

Benefits of Agile Project Management

Better product quality   Customer Satisfaction   More control on projects

Risk reduction   Faster ROI   37% Faster time to market   16% Higher productivity

The main principles of agile project management are quite straightforward, as you have already read in our previous chapter.

In real life, they translate into a variety of tactics and practices that make agile actually work. You can compare this with a car, to better visualize how this happens. If the Principles and the Manifesto are the theoretical laws of physics upon which agile is built, then the practices and specific tactics are the ways in which the different parts of a car are pulled together to make for a working vehicle.

This chapter is dedicated exclusively to teaching you *how* the Principles apply in practice, and how they can correlate with specific management tactics.

Let's jump in!

Customer Satisfaction

At the end of the day, customer satisfaction is why agile works, why agile was created, and why project management itself exists.

Customer satisfaction is the ultimate goal, the final frontier, and the pot at the end of the rainbow. It is what you aim for — you, your entire team, and your entire organization, in the end.

As such, every single one of the 12 Principles correlates to customer satisfaction. There are other benefits that come along the way, as pointed out in the first chapter of our book — such as the fact that you will make your life as a project manager easier and the fact that you will help your team members grow harmoniously.

However, when it comes to boiling everything down to one main concept, it is all about customer satisfaction.

While all of the Principles are equally important when it comes to delivering a product that will ultimately satisfy your customer or internal stakeholders, the first four focus more on making clients happy, while the others come as a kind of support in your endeavor.

To make sure you make your customers happy, you will first have to define who they are. In outsourcing companies, for example, this specific matter is quite clear: your customer is the company who ordered the software or product.

When it comes to developing internal products (such as, for example, an exclusively internal messaging tool), things might be a little fuzzy. In essence, however, the customer is the person or group of persons that ordered the product and with whom you will maintain continuous communication throughout the development of said product.

It is worth mentioning that, in general, project managers in agile don't necessarily talk directly to the customer — or, at least, not on a recurrent basis. The product owner is the person who takes on this role, ensuring proper communication between the customer and the project manager (who will later relay the information to the team and ensure all tasks are correctly assigned, timed, and financially managed).

The specific role of the product manager is to *translate* the customer wants into actual product requirements.

Say, for example, that your customer has ordered a social media management tool. They want to:

- Be able to post on Facebook
- Be able to post photos on Instagram
- Be able to schedule their posts
- Be able to monitor all their social media channels in one place

In product requirement terms, this might sound different:

- Cross-channel posting and scheduling

- In-tool image editing capabilities

The first four bullet points are the ideas the customer will come up with. The latter two, however, are what the product manager will relay to the project manager, who will in turn conduct team meetings and ensure that those product requirements are accurately split into micro-projects. Together with the team, the project manager will also make sure that working software is delivered on a recurrent basis.

How do you know your customer is happy?

Simply put, customers are happy when:

- Their needs are understood (the product manager's job)
- They see actual progress by receiving working software on a recurrent basis
- They see their feedback is actually followed and the product improves from one iteration to the next
- Their final product matches their initial needs, even if the requirements have changed throughout the duration of the development process.

It sounds both simple and complicated at the same time (especially if you have tried to maintain customer expectations and satisfaction at high levels before). Agile can help, though. The structure of agile is, in itself, one of the main ways customers in a variety of fields (including, but not limited to software development) are happy these days (as opposed to how they were, let's say, two decades ago).

Making and Managing Changes

A famous Greek philosopher once said that *change is the only thing constant in life.*

We constantly change, even when we don't actually *see* it happening.

We change from the very moment we begin life as embryos to the moment we fade out and return to Mother Nature.

It makes all the sense in the world, then, that agile would embrace change. In fact, this is one of the main tenets that makes agile so different from waterfall project management. While traditional project management methods see change as a boogieman waiting to attack from the back of the closet, agile takes change by its horns and embraces it.

Agile-based project managers understand that change is inevitable. They understand that customer requirements *change*, team structures *change*, and, ultimately, initial time-related and financial assumptions *change,* as well.

For those of you exclusively familiar with a traditional project management approach, the whole "embrace the change" concept might sound like downright chaos.

It isn't.

Same as waterfall, agile has its own processes and procedures to handle a variety of situations. The difference is that agile expects the unexpected — and, as such, it has developed its own way of dealing with change.

Although there are different ways of addressing all sorts of changes, the most universal method includes the following steps:

1. Understand the change

Sometimes, the change might come from the client (e.g. they have decided they need to integrate LinkedIn among the social media channels they want to manage via the tool they ordered from you). Other times, the change might come from the functional requirements (you have realized you need to implement a picture upload feature before you implement the actual scheduling tool).

At this stage, you should understand what the change means for the entire project and what its business purpose is.

2. Understand the scope of implementing the change

Regardless of where the change comes from, you have to understand its ramifications across all aspects of the project. Discuss it with your team and determine exactly what it will mean for your entire process, for the timeline you have created, and for the budget you had on paper.

Do make sure you take everything into consideration here and ensure your team agrees on all levels. It is important for everyone to be on board with the plan you are making, so that you can all implement it later on.

3. Ask for approval

This stage might be skipped under certain circumstances — such as when the customer specifically requests not to ask for their approval, for example, or when you know that your higher management has given you leeway in terms of approving such changes.

Other times, however, you will need to ask for approval of the change. In these cases, it is quite likely you will have to bring forward the implications of the change, as well as what you discussed with your team regarding what it will mean time-wise, resource-wise, and budget-wise.

4. Implement the change

When you receive your approval, you can proceed with the actual implementation of the change. Most likely, it will involve a re-planning of all the tasks and everything adjacent to them.

Changes can feel like little panic attacks when they occur. However, when you have an agile methodology to employ in your project management endeavors, you will find it much easier to handle all sorts of changes — from those related to the customer to those related to the cohesion of your team.

Continuous Customer Input

Agile projects are usually split into multiple iterations. They might be referred to differently from one agile method to the other (e.g. they are called "sprints" in Scrum, for example), but their main point is the same throughout the entire agile spectrum: to ensure that working software is continuously delivered to the client.

Receiving ongoing customer input is extremely important in agile because it ties back to providing customers with actual satisfaction when it comes to the final product.

In other words, you absolutely need your customer's feedback to ensure their happiness throughout the duration of the project — and upon the final delivery, as well.

There is a very good reason they call it *continuous* customer input. In agile, it is not enough to ask for feedback when you deliver a piece of working software at the end of a sprint. It is actually recommended to seek input *before* the end of the sprint. This way, each sprint will include the implementation of the feedback and the delivering of the final version of the working software.

Some customers might not be completely used to this MO. That's okay, you can help them understand that continuous feedback is much more productive than giant lumps of feedback that come sporadically. The more often you receive their input, the easier it will be to implement it — and, as a result, the happier they will be in the end.

Daily Meetings

Cooperation and collaboration are two of the core values in agile project management. Sure, ideally, waterfall project managers also want their teams to collaborate efficiently. However, traditional project management approaches rarely offer the actual *tools* to help the team coordinate, communicate, and cooperate effectively.

Daily meetings are fundamental when it comes to this. Not all project management approaches have this embedded in their systems "naturally," but most have borrowed the concept from Scrum project management, where the "Daily Standup" is almost sacred.

The Daily Standup is supposed to be a very short meeting where each team member takes two minutes to talk about:

- What they did the previous day
- What they plan on doing today
- What are their bottlenecks/roadblocks

This type of meeting might sound like micromanagement to some, but its purpose is completely different. Scrum Masters and agile PMs don't hold the Daily Standup with the purpose of following each team member's every move.

On the contrary, the main purpose of the daily meeting is to allow team members to coordinate with each other and enable the PM to do their job: removing bottlenecks and roadblocks as necessary.

The Daily Standup is also a great exercise in honesty and self-discipline. When everyone takes their turn to stand up and actually be honest about what they managed to accomplish the day before, you will create a spirit of self-organization. People will feel more accountable for their actions (or lack of). And, the more this ambiance stretches throughout your entire team, the more they will *naturally* flock towards agile project management (and love it!).

The daily meeting might also sound like a waste of time and a potential disruption. However, it creates a kind of routine that puts people in a productive mood and helps streamline the entire process. As such, instead of being a disruption, it is a small chunk of your workday you are investing among the remaining seven or eight hours of work. In other words, you might spend 20 minutes for the entire meeting — but the results will definitely pay off, inevitably!

Sustainable Development

A lot of people mistake the idea of *agile* development with the idea of *fast* development.

Yes, agile projects tend to be delivered faster. But that's due to the way in which they are split, managed, and coordinated with all the changes that might come along.

Agile projects are not delivered faster because the PM pushes the team over their realistic limits to deliver *faster, faster, faster*.

On the contrary, actually. In agile project management, you have to make sure that you support a sustainable development approach — the kind that doesn't push team members over their limits, doesn't overwork them, and doesn't create unbearable stress in their lives.

It sounds like a dream (for pretty much everyone, from product managers to project managers and developers alike), but it is feasible.

Do keep in mind that it is very easy to fall off the wagon and start pushing for *more* in *less time*. It is in the nature of the alert rhythm of agile projects to be tempted to ask for more than what your team can deliver.

However, as long as you can balance out this instinct, you will be just fine. In fact, you and your team will discover that people tend to be more productive when they have plenty of work to do, but aren't constantly stressed out.

You don't have to take our word on this. According to studies (Astutis, 2019), stress can actually impact your team on multiple levels:

- They will be more prone to take sick days
- They will be more prone to quit the job and move elsewhere
- They will be more prone to be less productive during work hours
- They will be more prone to deliver low-quality work
- They will be more prone to ruining their work relationships

In other words, stress can be a real monster you have to vanquish from your team. And one of the main ways to do this is by ensuring that you plan your project in a very sustainable way, from the very beginning.

It is important to talk to your team and discuss exactly how they think the tasks should be prioritized, how long each one will take, and whether or not they think some tasks won't be able to fit into one iteration only. They know their jobs better than anyone else, and they can make accurate estimations on what they can and cannot do.

Trust them. This might feel odd at first, specifically if you are used to a traditional management approach where you don't involve your team as much. However, it will pay off — when people feel trusted, they are less likely to be dishonest by either overestimating their own abilities or underestimating them.

Continuous Improvement

Task management in agile is built around the concept of iterations. Once the project has been split into multiple stories and/or epics, you will want to make sure each user story can be done within one iteration, and that epics will have a recurrent cadence throughout the iterations they will cover.

The whole point of splitting projects into multiple iterations/sprints is to make sure the following things happen:

- The client receives working software on a constant basis
- The client can provide regular feedback throughout the course of each iteration and at its end
- The client can ask for changes if needed
- The project is split into smaller chunks that are easier to manage
- The risks associated with the project are more closely monitored
- The product is continuously improved as a result of all of the above

Repeated and well-cadenced iterations are crucial because they will allow you to create a sense of continuous improvement. This will be reflected on at least two levels:

- That of the product (ergo, customer satisfaction)

- That of the team (as a whole and for each individual, as well)

We have already discussed that customer satisfaction is the blood that runs through agile veins. We do want to take a moment to discuss the *team* level too, though.

Continuous improvement has to reflect on each individual in your team. Each of them has to grow and become better, from one iteration to the other. For this purpose, you will have to learn how to be a leader, rather than a distant project manager.

We will approach this matter in more detail in our last chapter in this book, but for now, it is important to keep in mind that continuous improvement doesn't refer only to the actual product — it needs to expand across the entire team and the way they perceive their work.

Simplicity as a Vital Element

Most of the agile project management methods put quite a lot of emphasis on the idea of simplicity (it is, after all, one of the main Principles).

In Kanban, however, simplicity becomes the very essence of being. Every single agile concept connected to simplicity is shrunk down to its very basics. For instance, in Kanban, team members will not be allowed to handle more than one task at once. The Kanban board is a simplified version of the Scrum board and, in the end, the idea of simplicity runs deep into everything Kanban projects are.

In other agile project management methodologies, simplicity is a more general (but nonetheless important) concept.

Most of the time, it applies to the idea of removing all obstacles, including (but not limited to) tasks that might not be needed.

Going back to our cake example, would you actually *need* to add star-shaped glitter sprinkles on top of the frosting once the cake is ready?

Probably not. It's something you can do, but it is not a must for the cake to be edible (or delicious, for that matter).

The same goes with agile tasks, as well. If it doesn't add real value to the final piece of workable software you deliver at the end of your sprint, then it can be easily removed from your list of to-dos. What is the point of leaving it there? It can become a distraction, it can stray you away from your path, and it can completely ruin your sprint, in the end.

Keep it simple — as simple as you can. If you have to strip tasks down to their bare bones, then so be it. You might be surprised just how far a little simplicity can take you!

Self-Organizing Team

As mentioned earlier, the idea of micromanagement is completely incompatible to agile project management.

Your goal as a project manager is not to ghostly hover over everyone's work and make sure they don't scroll their Facebook homepages for more than two minutes a day.

Your goal is to build the kind of environment and work atmosphere that will make people actually want to work on their own, without feeling the threat of punishment if they have a bad day and without feeling constantly pressured by the stressed expressions of their bosses.

Your goal is to keep your interventions to a minimum:

- To simply act as the Master of Ceremonies during the daily meeting
- To help remove roadblocks for team members
- To intermediate the communication between the product manager (and what the customer tells them) and your team (as well as the other way around)

You don't want to be the kind of project manager who follows what each team member does every single minute of the day. Not only does that show a severe lack of trust (and creates frustrations and animosities), but it is also counterproductive.

Yes, there may be times when you will have to have one-on-one chats with your team members and help them work more efficiently. And yes, there may be times when you will feel like you *need* to closely watch over your team's minutes at work.

You have to fight off those instincts, though. Ideally, your team will fall into place, eventually. Their composition and structure will also settle at some point. And sooner or later, they will not only *get used* to agile project management (and all its quirks), but they will start to genuinely prefer it, as well.

Inspect and Adapt

As we reiterated before, continuous improvement is a fundamental concept in agile project management.

In fact, it is so fundamental that agile has made it part of the whole process, every step of the way.

Inspecting and adapting is one of the most essential aspects of agile project management — and it tightly correlates with the 12 Principles of agile project management. Without this, you literally have no agile.

To reach your goal of constantly improving (the product and the team), you will have to also make sure you are very consistent about inspecting and adapting the product.

There are different approaches as to how often this should be done. However, most agile methodologies agree that inspection and adaptation should be performed on a regular basis.

Furthermore, do keep in mind that the Inspect & Adapt workshop (I&A workshop) is different than the bi-weekly sprint review analysis. In general, the I&A workshop consists of three main stages:

- The PI System Demo (where the working software is demonstrated)
- Quantitative Measurement (where specific pre-agreed metrics are measured)
- Retrospective (where conclusions are drawn for future implementation)

The I&A workshop is the key to ensuring constant, consistent, and continuous improvement when it comes to the product. It is also the key to helping your team grow and making your customers happy — which circles us back to the ultimate golden rule of agile: client satisfaction.

A lot that can be said about each of the 12 Principles and how they translate into real-life actions and projects that are genuinely successful. We chose to brush over the main concepts, so that you can gain a better understanding of how agile functions and how it can help your organization efficiently deliver better products that are more in line with your budget and, ultimately, more in line with each of your team members' personal development goals.

Chapter 3: Implementing Agile: How to Apply the APM Method Effectively

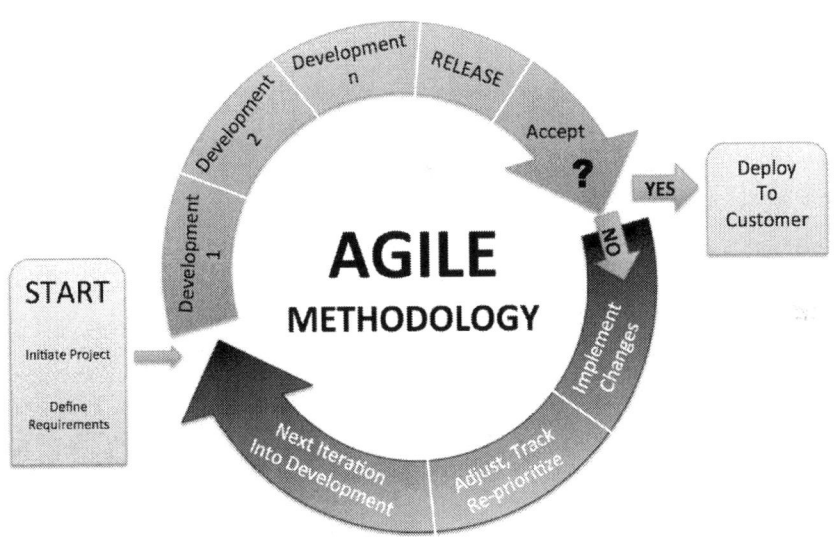

In theory, agile project management sounds really easy. It all boils down to:

- Making sure all product requirements are understood
- Making sure product requirements are translated into stories
- Making sure each story is split into the appropriate number of tasks
- Making sure each task is properly prioritized
- Making sure you take the iterative process as an opportunity for continuous improvement

- Making sure working software is delivered at the end of each sprint

Doesn't sound so bad. Given that there is less emphasis put on documentation, it might even sound a lot more effortless to some of you.

However, what you do need to understand (and what we definitely need to be honest about) is that agile is not easy. Project management in general is not easy, regardless of how traditional or agile you decide to be.

It is quite important to acknowledge the fact that randomly applying the Agile Principles and Manifesto *might* help. But if you want real results, you should do so consistently, and you should make use of all the tools and techniques agile provides you.

This chapter is all about exploring the specific stages of implementing agile the *right* way. Don't get this wrong: there is no recipe to follow, as we have already mentioned. However, the specific techniques we will discuss in this chapter are universally accepted as standard in agile, and they should be considered when you start implementing it as a new project management approach.

Indeed, agile project management is all about being flexible — in terms of tasks, product requirements, changes, and the method itself. There is a reason there are so many types of "purebred" agile and hybrid agile methods: every organization has different needs, and agile project management understands that. As such, it has developed its own universal mechanisms while leaving plenty of leeway for adjustments as per the organization's needs, protocols, and goals.

The techniques and bits of advice we want to present in this chapter will be equally useful regardless of which specific agile methodology and framework you might choose. Be it Scrum, Kanban, Lean, or any other agile methodology, you can rest assured that these tips will *work* at a universal level.

Define Your Vision

This is where it all begins, where the entire project is defined in terms of what you want to achieve with it. Of all the stages in the development of a project, this one will always feel the most motivational and inspiring.

Defining your vision is not about empty words, though — and you should definitely pay attention to this. Yes, it can be an inspiration and yes, it can definitely get quite creative, depending on the type of product you have to develop.

Furthermore, it is also important to mention that this is not a step to skip. No matter what type of agile project management methodology you might choose to follow, it is essential to take your time at the beginning of the project and define the vision.

Why is this stage so important?

Simply put, it will outline everything you will aim for throughout the duration of the project. In some ways, you can look at your vision as the lighthouse that will guide you through the product development process. Whenever things go south, whenever you and your team get lost along the way, whenever you feel that it will never end, your vision will be there to lead you back to the right path.

Your vision doesn't have to be something larger than life. In fact, your vision definition should be quite succinct — not in the sense that you should speed up the process, but in the sense that you don't have to write an entire novella about the product you are planning.

There are multiple ways to tackle the vision definition stage of your agile project development. One of the simplest is to simply answer a handful of basic questions:

- Who is the product for?
- What do these people need or want?
- What is the product category?
- What are the key benefits the product brings forward?
- Who is the product's main competitor?
- What does the product do differently than its main competition?

When you answer all these questions, you will be able to write down your vision. For instance:

"AgileSocial is a social media management tool aimed at large agencies who need to manage tens of accounts in one place. The main benefit of AgileSocial is that it provides custom integration with more than just the basic social media channels (Facebook, Instagram, Twitter, and LinkedIn). Unlike Hootsuite (competitor), AgileSocial will provide more features in the free plan, such as the possibility to manage 20 accounts."

This basic statement can do the trick, especially when you are developing your own product and you don't have a large team to work with. However, if you want to spend a little more time defining your product vision, you should consider the following steps:

1. Ask all the questions we have previously mentioned and draft the basic product vision, as per the example above.

2. Validate your vision by asking yourself a new series of questions:
   - Is your statement clear?
   - Is your statement meant to be read by your team and other internal stakeholders?
   - Is the description comprehensive and compelling enough when it comes to explaining the customer needs?
   - Is the description describing the very best outcome the product development could lead to?
   - Is the business objective clear?
   - Is the vision congruent with the organization's vision and values?
3. Validate your vision with the following people:
   - Internal stakeholders
   - Development team
   - Scrum Master (in case you aren't the Scrum Master)
4. Rethink and rewrite your vision as per the input you have received after implementing the second and third points in this list.

Regardless of whether you want to take the long way or the short one, this is a stage you simply shouldn't rush past. It might not sound like much, but you have no idea just how lost you can get when change occurs and your initial plans are turned upside down. In these situations, having a clear written vision statement can be a real life saver from many points of view.

Life Cycle Selection

Everything in our Universe has a life cycle. In fact, the Universe itself has its own life cycle, as well — one we might not be able to decipher just yet, but one that has been theorized in a number of ways across various branches of physics.

Your project has a life cycle of its own, too. Defining this life cycle is important, because it will give you a clearer blueprint on what you should do throughout the entire duration of the project — the techniques you will employ, how tasks will be assigned, and how everything will be delivered in the end.

In project management, there are several types of life cycles acknowledged:

The Predictive Life Cycle

This type of life cycle is frequently associated with waterfall project management because it is focused on having a very clear, very minutious plan ahead of the project development. From the very beginning of the life cycle, you will know the specific scope of the project and you will be able to estimate the duration of the project, as well as the costs associated with it.

It's not that predictive life cycles are inherently bad (just like waterfall project management isn't bad in itself). It's just that this type of life cycle is actually pretty hard to follow, unless the product is very, very well understood. For instance, if your project is all about building a new version of an existing car model, you might be able to use the waterfall approach and opt for a predictive life cycle.

If, however, there is even the slightest doubt to shadow the carefully planned blueprint, you might want to consider a different type of life cycle. More often than not, both requirements and conditions change throughout the product development process — and this means that you likely won't be able to follow the strict plan you had in mind. Therefore, both delivery time and cost expectations will be ruined, both on the side of the organization you work for and on the side of the customer.

The Iterative Life Cycle

This type of life cycle is associated with a primordial agile project management method. According to the iterative life cycle, you will progressively change and improve the product, step by step.

Iterative life cycles are very tightly connected to the idea of going into greater detail every step of the way. The more details you can pan out, the more you will be able to improve the product at a granular level.

Iterative life cycles are usually applied in two main circumstances:

- When the product requirements are not yet very clear
- When the plan you have created has unclear requirements

With this type of life cycle, you cannot expect to have the full scope of each iteration prior to the actual beginning of each iteration. This is largely due to the uncertainty of the project itself and the iterative nature of the product development.

In other words, iterative life cycles will allow you to discover the product every step of the way, just like you would do when peeling an onion. Stage by stage, you will gain more details on what your product will look like and what is actually needed — and you will adapt from one iteration to the other.

The Incremental Life Cycle

Incremental life cycles deal with projects that are very complex and use technologies and platforms that are not yet fully developed.

Normally, you wouldn't even begin the development of a project until you have at least a set of basic requirements and a working platform from which you can build. However, when customers want their products to be delivered very quickly, you might be forced into adopting an incremental life cycle.

It is worth noting that most of the incrementally-developed projects do not focus on aesthetics, and often, they don't even focus on UX design (where "UX" stands for "User Experience" and defines a design approach focused on making things easy for the end user).

Usually, incremental life cycles are used for projects where *timely* delivery is the most important factor. Given the circumstances, it will happen quite often that these projects will be delivered as an unpolished draft, rather than a fully-usable product.

Obviously, you can't apply this life cycle to products that need to be functional from every point of view. You can apply it to functional software, for example, that doesn't need to be actually *pretty*, but simply needs to run a set of basic features.

The Adaptive Life Cycle

In a world where agile is a completely balanced approach (not traditional, as in the case of the predictive life cycle, and not extreme, as in the case of the iterative and incremental life cycles), the adaptive life cycle is the closest to the agile ideal.

The adaptive life cycle combines the iterative and incremental life cycles to create a uniform, balanced, and well-rounded approach to agile project management.

Imagine the adaptive life cycle as a game of building blocks. Stacking one block on top of each other means that you are building everything in an incremental way. Giving your building to the customer at the end of the sprint and starting all over again to improve the building means you are taking both an incremental and iterative approach.

With agile, you deliver working software at the end of each sprint, and you start again with every new sprint.

As shown in this book, the combination between an iterative and an incremental approach will lead to products that are delivered faster, better, and with cost efficiency in mind.

There is no right or wrong when it comes to life cycles — but you do have to make sure you select the one that best suits your particular situation. To reiterate and refresh:

- The predictive life cycle is all about projects that are highly predictable
- The iterative life cycle is all about projects that offer you a decent level of knowledge in terms of product requirements, but are still not very clear
- The incremental life cycle is all about projects that need to be developed on fast-forward
- The adaptive life cycle is all about agile projects that provide you with plenty of information to begin with and allow you to continuously improve the product, one iteration after the other.

Analyze your specific situation and choose the type of life cycle that suits you best!

Creating an Agile Environment

Beyond all the tactics and the specific tools used by each agile project management approach, there is one thing that lingers in the background to support agile not only in a technical way, but in an ideological way, too.

Why is this needed?

Because agile project management is not about spreadsheets and formulae, it is about people. And when it comes to people, you need to do everything in your power to ensure that they feel comfortable within the spectrum of agile, in your organization.

Creating the right agile environment will help you and your team connect, collaborate, and foster an ambiance of complete honesty. When these requirements are met, teams are likely to perform much better, they are likely to make better estimations, and they are likely to deliver better products in less time.

The so-called "agile environment" is not about the physical environment only, but the organizational culture and policies that support the entire company.

There are many things you can do to create an agile environment for yourself and for your team members. However, the most important points include the following:

Setting the Example

Truth be told, you cannot expect people to automatically jump aboard the agile train. In all honesty, you might have to deal with a couple of team members who will be completely skeptical about the entire concept.

That's okay, and it's perfectly normal. People are reluctant to change, and even more so when it challenges the very way they have been doing their work. Add to this the daily meetings, and you can easily see why some people might not be *that* happy about the adoption of agile.

What you need to do, first and foremost, is set an example. Show honesty and open-mindedness, listen to your team's concerns, and prioritize discussing things with them. Be the change you want to see in your team!

Being a Leader

Agile project management doesn't deal in terms of "bosses" and "employees." Nobody's a *"boss"* in agile. Many people perceive the project management role as a bossy type — but in agile, this is completely untrue.

Instead of bossing people around, be a leader. Be the kind of PM that helps each team member grow at an individual level, so that they can help the team move forward. Be the kind of leader who inspires!

Cultivate your leadership skills at each opportunity and you will create an environment that promotes collaboration over ranks and improvement over stagnation.

Adopting Critical Thinking

A good agile project manager needs to possess the gift of critical thinking. Of course, you have to be understanding about the changes that might occur in each team member, and you have to listen to their needs, desires, and fears.

However, when it comes down to it, critical thinking will help you make the difference between being a project manager who just *lets things be* and a project manager who is downright negligent.

Apply your critical thinking to all verticals of project management, not just the way you manage your team. Use it when you analyze tasks, the time it takes to do each one, and whether or not the customer requirements are realistic when compared to the team's capabilities and the given budget.

Stimulating Collaboration

We have already mentioned this a number of times: collaboration is quintessential in agile project management. Without it, you cannot actually work as a team, you cannot stick to the iterations you plan, and you cannot deliver on time and within the budget.

Always stimulate collaboration between your team members, between yourself and your team, between the various stakeholders in the company and your team, and, ultimately, between your customer and your team (via the product manager, if there is a separate person taking this role).

Plan "Just Enough"

This taps into the basic Principles of agile project management and it focuses on *not* doing extra work.

In traditional environments, extra work is always well-seen.

In agile, however, it is a waste of time and resources.

This is not to say that you and your team shouldn't be proactive, but you should focus on delivering the tasks at hand first, *then* show off your bigger thinking by adding something on top.

Plan *just enough* to make it through each iteration, don't go overboard. It's not a lazy approach, it is a focused one!

Encouraging Face-to-Face Conversations

As we have already mentioned, face-to-face collaboration and communication are always encouraged in agile project management.

It is worth noting that, over time, modern technology has allowed remote teams to work in a much smoother way. These days, you can have daily video calls for free — and they can be a good replacement for actual face-to-face conversations.

At the same time, we also want to mention that you should promote business trips as often as you can. At the end of the day, nothing can compare to being *there*, physically present with your team members and having genuine discussions with them.

Creating the Physical Agile Environment

The physical environment can have a pretty big impact on productivity — and, as such, agile project management suggests that you create offices that enable people to communicate better.

Open-space offices where everyone can discuss freely are quite essential in agile. At the same time, you should also make sure to have a few meeting rooms available (including a formal and an informal one), spaces where people can work in silence, and spaces where people can have personal phone calls, too.

All of these things can complement the productive environment you are trying to build, promoting communication and collaboration between all of your team members and helping them be the best version of themselves.

Delivery in an Agile Environment

Since agile project management focuses on an adaptive life cycle that combines both the iterative and the incremental building of the product, it makes sense that delivery will be *adaptive,* as well.

In essence, you have to deliver working software with each sprint/iteration. This means you might deliver a new feature with each sprint, or that you might improve on features that have already been built.

Delivery should always be followed by customer feedback and by the implementation of said feedback.

Obviously, your main focus is to make sure working software is continuously delivered and that the customer is always satisfied with the results of your sprint's work.

However, you should also keep in mind that constantly talking about deadlines and deliveries might put a lot of pressure on your team members. They are all adults and very much aware of the fact that your organization has taken the responsibility of delivering a good product at the end of the project. There's no need to continually remind them that they have deadlines to adhere to.

In fact, agile project management goes against this completely by allowing people to take moments of relaxation during the workday, as well. This is precisely why you will see ping-pong tables, relaxation areas, and even game rooms in many software developments companies — these businesses want their workers to be happy and productive, and they will go to great extents to make this happen.

Dwelling on time and pressuring the team might also make them feel less involved in the project every time they are reminded that they are running against the clock. Instead of focusing on minutes, hours, and days, start looking at how you can make your team feel truly involved in the development process. Talk about value, rather than time — it can make all the difference in the world.

As for *how exactly* each bit of the project should be delivered, it all depends on you, your team, and your customers. Collaborate and agree on terms of delivery that are feasible and sustainable — it will only make for a win-win situation in the end.

Organizational Consideration for Project Agility

On paper, agile project management might sound like a dream come true from every single point of view:

- That of the PM, who will finally be able to *manage* and *lead*, rather than bury their head among the hundreds of columns of an endless spreadsheet;
- That of the team, who will learn how to self-organize and how to genuinely grow without a constant Big Brother to watch over their shoulders;
- And that of the customer, who will receive a better product in less time.

But the truth is, if you haven't yet adopted agile, it is most likely because your organization is quite traditional in the way it *does* things.

As such, your agile deployment plan has to include the organization. You shouldn't expect to move mountains and convince C-level executives that your little games are actually helping developers be more efficient.

Instead, you should be realistic.

And you should make a proper business case for the deployment of agile — if not across the entire organization, then at least within your team.

There are plenty of resources to help you prove that agile project management shows actual results. There are tons of courses your team members can take (including courses you can develop to train them in the art of being agile!). And there are tons of *massive* companies you can use as an example, too.

You might not be able to change your entire organization. However, if you are able to get a few higher managers on board with you, you will definitely make a grand step. Do keep in mind that they have to not only agree to a new project management approach, but they will have to actively support collaboration and communication, as well.

Moreover, if your company focuses a lot on documentation and procedures, it is necessary for you to show them that you won't drop the "old ways" entirely — you will just give them a more modern, more flexible spin, to improve your team's efficiency and delivery timings.

Call to Action

Agile project management is not the kind of thing you think up one morning and simply bring to work.

It's not chocolate, or the cake you baked last night.

It's an endeavor, a constant process of improvement and adaptation, and, ultimately, a project within itself.

Deploying agile project management takes hard work and a deep understanding of how this method works, how your organization works, and, essentially, how *humans* work.

This is precisely why it is very important to construct a call to action that supports your specific situation. As mentioned in the previous section, there are many resources you can use to make a business case for the use of agile project management. However, if you don't adapt all these statistics and examples to your company and its situation, you risk sitting in front of an unimpressed board of directors who will just dismiss agile as the *new trend* all the kids want to try out.

That might be an exaggeration, but we made it with a point. Put yourself in the position of a decision-maker in your company. Would you really want to implement something as disruptive as agile project management if you didn't hear clear arguments that it will actually work for the very specific case of your organization?

Probably not.

That is precisely why you need to take your time in making your business case and call to action. Even more, put the same amount of effort in presenting your agile ideas to the team members, as well. We will discuss more on how to build the perfect agile team towards the end of this book, but until then, you need to make sure people are on board with the change — without them, you simply cannot lay the foundations of agile in your company.

Chapter 4: Tools and Methodologies for Quality Control in APM

Although agile project management only became official two decades ago, it has rapidly advanced to a full-blown approach. As such, it has started developing its own tools, as well.

This chapter is dedicated to helping you pick the best tools for your agile organization and team. Each of the tools we will mention here is well-regarded, so it cannot be said that one is necessarily better than the other. They are different in many ways and they tap into different agile project management needs — it is more than worth analyzing all of them if you want to make an informed decision.

Clarizen

This is one of the most popular project management tools at the moment. Its main features include workflow management, collaboration facilitation, task management, resource management, financial management, and integration features.

Clarizen comes in two versions: Enterprise and Unlimited. Both will allow you to access pretty much the same features, but the main difference between them is that the Unlimited version will allow more customization. Both versions are available with free trials, as well, so you have the opportunity to test them out.

In its essence, Clarizen is quite inexpensive (especially when compared to other project management tools), so it is definitely worth giving it a try.

Trello

This is by far one of the most commonly-used project management tools. It is frequently used for Kanban projects because of its column-based interface, but you can use it for any type of project, really.

Trello is free to use — and it is quite fun to use, too. It will allow you to create columns for each of the steps in the workflow (Backlog, Doing, Pending, Done). Once each task is completed, it can be moved into the next column (you just have to drag and drop each card).

GitScrum

This tool is based on GitHub, one of the most famous version control tools out there, and its main purpose is that of seamlessly integrating a Scrum project management tool with your version control tool.

GitScrum has four versions: Free, Freelancer, Business, and Professional. Of these, the Business version offers the most features (and it is the most expensive option, too). The Free option is a decent version to try out, especially if you want to test out the main features of this project management tool before committing to it.

Jira

If you visit any software development company and take a look at their tools, it is more than likely that you will find Jira to be one of those they use on a daily basis.

Jira is, without a doubt, one of the most famous project management tools on the market at this point. As a ticket-based/issue-based tool, Jira will allow you to easily track all issues that pop up in the development process and alter it accordingly.

There are two main versions of Jira: Standard and Premium. Both can be tried for free for 30 days. Moreover, it is worth mentioning that Jira can provide you with enterprise solutions, if that is what you're looking for.

Taiga

In terms of complexity, Taiga is almost as large and as extensive as Jira. The one differentiator? Taiga is an open-source project. This means that it is developed for free by programmers who want to participate in the project and bring improvements to the software.

Taiga is free to use in its Basic version and it costs a meager $5/month for the Premium version. There is also an Enterprise version, but additional information can only be accessed by reaching out directly to the company.

Nostromo

Of all the tools we have picked for this chapter of our book, Nostromo must win the prize for the most unique design. The entire tool is centered on the concept of space exploration, and even its versions are named according to this theme: Satellite Sparkle, Starfield Firefly, Comet Teddy, and Stardust Giant. Prices start at $10/month, and they go up to $100/month for the most advanced version of the tool.

For the money, you are getting a project management and product management tool that brings everything in one (very well-designed!) space: tasks, reports, time management, and so on.

Hansoft

Like the other tools we have mentioned, Handsoft focuses on bringing as many features as possible to one place. The difference here is that Hansoft is built to be an enterprise tool from the very beginning. Ergo, it offers enterprise-specific features embedded in its default version —

such as the ability to see the backlog for the entire organization, or features related to bringing remote teams closer.

Furthermore, Hansoft will allow you to work with multiple project management approaches — including Scrum, Kanban, waterfall, and even mixed methods.

This tool is free to try for up to five users. Contact the company for more information if you decide to get it for your organization.

Blossom

As the tagline suggests ("Project tracking for distributed companies"), Blossom is meant to be used by distributed companies. Its features focus on making things as clear as possible for the project manager — including the way he or she understands the roadblocks in the development of the product.

Blossom has very good integration with GitHub, and it can also be integrated with other communication tools (like Slack and HipChat, for example).

For pricing, you will have to contact the company directly.

Ravetree

This tool is excellent for agencies, for creative teams, and for project-driven companies. The features focus on rolling all the tools together in one place: from the insights and metrics to the communication, and from the remote teams to resource organization.

Ravetree has three payment options for the same version of the product: for those who want to pay annually, for those who want to pay quarterly, and for those who want to pay monthly. If you want to pay once per year, it will be $29/month/user. If you want to pay monthly, it will be $39/month/user.

Of course, these are just some of the project management tools out there. Each has its own specificities, its own advantages, and its own disadvantages — analyze them comparatively, try out those that seem a better fit, and settle on that which makes your job and the job of your team members easier and more productive.

Chapter 5: Managing Risk in Agile Project Management

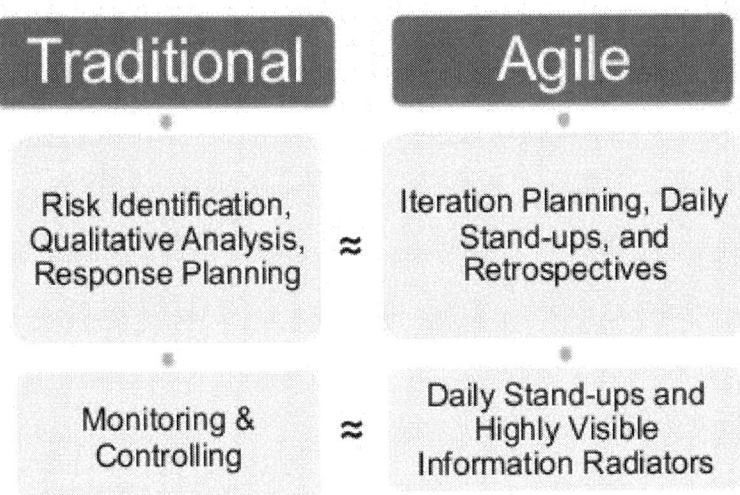

Risk management is a quintessential element of *all* types of project management. It is, ultimately, where real project management starts and where great project managers can show off their true value.

Managing risk in agile project management is tackled a bit differently than it is in the more traditional approaches. In waterfall, risk management is done based on the mega-plan created at the beginning of the project.

In agile, however, risk management is done iteratively, with each new release. The more sprints you go through, the more accurate a view you will have on what the risks are and what you can do to avoid them.

This is not to say that you will jump in headfirst when you start your first few iterations. However, you won't be able to (and you shouldn't) plan down to the smallest detail. In agile project management, this is considered to be anything but productive, specifically because the nature of the requirements and of the project itself can change — and, as such, the risks will change, as well.

For those of you who have only worked with traditional project management methods, agile risk management might sound like complete chaos.

However, it is just as organized and just as structured as any other project management method.

We will dedicate this chapter to teaching you the ins and outs of risk management in agile project management. While we may not dedicate as much space to any other individual aspect of agile project management, we feel that risk management is too crucial, too large, and too misunderstood. As such, we want to give it plenty of time and space, so that you fully understand how it is done.

Classify

Identifying risks in agile projects is all about discussing stories with your team and finding those parts that might prove troublesome along the road. The risk identification usually takes place in the third stage of the project development (once you have already understood the product requirements and the size of the stories).

The reason risk should be managed during the third stage is because by then, you will have a deeper understanding of the type of project you are dealing with and a specific idea of what you have to do.

Once identified, risks should be classified according to their nature. This step is important because it will help you see the risks you are facing more clearly — allowing you to plan preemptive actions to help you avoid those specific pitfalls.

Of course, since this is agile project management, you should accept the fact that risks might change over the course of the project development itself. This is precisely why risk management should be a part of each iteration planning — the more granular you can go, the more specific the risks you identify can be.

There are multiple ways to classify risks in agile project management, but we will discuss two of the most popular ones.

The Descriptive Method

This risk classification method is not prescriptive in the sense that it will allow you to alter the risk classes as you see fit. Some projects may meet all risk classes, others might be more limited (and, as such, they will only have to deal with specific classes of risks).

Some of the most important and common classes of risks you might encounter include the following:

- Solution — risks that affect whether or not the final product is an actual solution to your customer's needs.
- Timeline — risks that affect whether or not the project will be delivered on time.
- Budget — risks that affect whether or not the project will actually stick to its initial budget planning.
- Privacy — risks that affect whether or not the project can comply and abide to privacy regulations and legislation (such as GDPR, for example).
- Security — risks that affect whether or not the project and its contents are safe from potential hackers.
- Resources — risks correlated with whether or not you will have enough resources (human and non-human) to finish the project.
- Scope — risks that affect whether or not the project is contained correctly.

Depending on the type of project you are managing and its very specific implications, you might also want to take into consideration political, environmental, or reputational risks, for example.

Also, it is important to note that some risks might fall in multiple classes. For instance, lack of senior developers can be a resource-related risk, but it can also be a budget-related risk (as you will have to spend more money for good developers) and a time-related risk (as you will have to allocate time to source and recruit new developers).

The PESTLE Method

There isn't much to differentiate what we called the "descriptive" method and the PESTLE method, aside from the fact that the latter uses an acronym to make it easier for users to remember the main categories of risks to consider in the planning process.

PESTLE stands for:

- Political
- Environmental
- Social
- Technological
- Legal
- Economic

As you can see, these risks are more general and extrinsic to the project itself, so the method might be more suitable for projects that are actually related to the political, environmental, or social scene (but not exclusively so).

Same as with the other method, some risks might fall in multiple classes — it is important to discuss matters with your team and analyze these risks from every imaginable perspective.

You can't predict the future, yes. But you can definitely anticipate scenarios — and this is where risk identification and classification should occur.

Quantify

Identifying risks and classifying them according to their nature is obviously important. At the same time, it is also worth mentioning that quantifying each risk according to its level of importance and the impact it can have on the project is also crucial.

You may or may not like this, but numbers give meaning in project management — and although agile might be a more innovative way of dealing with the management of projects in general, it still abides to the same old-fashioned rules in terms of assigning numerical importance to various elements.

Same as classification, risk quantification can be done in multiple ways. Because it is relatively simple and because it can be just as good as any other technique, we have chosen to present you a method that uses a basic matrix to help you identify risk importance.

Basically, this method uses a matrix made up two axes: Impact and Probability. Each risk will be assigned a certain numerical value for the Impact axis and a certain numerical value for the Probability axis. When the two axes are brought together, each risk will be associated with a number that reveals just how critical that specific risk is to the project as a whole, and whether or not it should be tackled first.

When assigning numerical values for the Impact axis, follow this guide:

- 1) Minimal — the impact it will have on its class(es) is minimal and the risk impact should be reviewed every three months.
- 2) Nominal — the impact it will have on its class(es) does not exceed 5% (e.g. the project budget will be overrun by 5% if this risk occurs).

- 3) Moderate — the impact it will have on its class(es) can be somewhat significant, but the occurrence is unlikely and it will not affect the project by more than 10 percentage points (e.g. the project budget will be overrun by 10%).
- 4) High — the impact it will have on its class(es) can be significant and it can affect the project by 25% (e.g. the project budget will be overrun by 25%).
- 5) Extreme — the impact it will have on its class(es) is major and it can affect the project by 50 percentage points (e.g. the project budget will be overrun by 50%).

When assigning numerical value for the Probability axis, consider the following:

- 1) 0-10% — quite unlikely to occur
- 2) 11-40% — unlikely to occur
- 3) 41-60% — may occur
- 4) 61-90% — likely to occur
- 5) 91-100% — very likely to occur

Once the matrix is filled in, you will multiply the numerical values assigned to each axis and get a number. Starting with that number (called Risk Value), you will be able to plan ahead and correctly prioritize the way in which you plan ahead to avoid major risks and to be as certain as possible that they will not affect the smooth progression of your project.

Plan

The Risk Value associated with each risk will give you a clearer idea of just how important it is to tackle that risk as soon as possible.

In general, there are four major categories of risks:

- Minimal (associated with Risk Values between 1 and 5)
- Moderate (associated with Risk Values between 6 and 12)
- Serious (associated with Risk Values between 12 and 20)
- Critical (associated with Risk Values between 20 and 25)

Different types of planning will be required, according to the category of risk each situation falls in. For instance:

- Minimal Risk Values do not require any kind of specific action, as their impact is slight. They should be reviewed quarterly, as mentioned before.
- Moderate Risk Values do not require actual action, but a senior manager should be notified of them. Moreover, they should be monitored and reviewed on a monthly basis.
- Serious Risk Values will require you to notify a senior manager and to review and monitor these risks on a weekly basis.
- Critical Risk Values require immediate action. They should involve executives, and they should be monitored on a daily basis.

Examples of risks to include in each of these Risk Value categories can be as follows:

- Minimal: The computers the developers are working on are starting to age.

- Moderate: Multiple team members will be affected by a bout of flu.
- Serious: The budget is stretched out too close to its upper margin.
- Critical: The system is hacked and information has been spilled out.

As you can see, everything in agile can be just as mathematical as in any other project management approach. The main difference between agile and waterfall here is that agile will encourage you to re-assess risks on a regular basis, depending on how important they are.

In waterfall, all these risks would be panned out at the beginning of the project and they would not be re-assessed under normal conditions. This means that new risks might pop up along the way and already identified risks might escalate in importance, and you would not be ready to face them.

Act

This stage of project management involves taking actual action against the risks you have identified, classified, and planned for. Action should be taken as per the plan, and it should include specific steps that work towards minimizing or completely eliminating the risks, one by one.

Since the action itself is highly dependent on a multitude of factors (importance and nature of the risk being the most essential ones), we won't dwell too much on this. It's all about the ways in which you work with your team to eliminate as much risk as possible, so you can stick to your project plan and what each iteration is aiming to deliver.

Repeat

The incremental and iterative nature of agile project management manifests in the way you run your risk management, as well.

The entire risk management process is another way in which you apply the Assessment and Action cycle — just like you do with the development.

How often should you repeat the agile risk management cycle? In general, it is recommended to do so at least quarterly. But the very best way to handle it is by tackling risk-related questions and answers with each sprint planning. This will allow you to have a more granular control over the risks your project is facing.

One concept you should definitely keep in mind is that of the "early failure." The earlier in the process failures occur, the easier it will be for you to identify solutions before risks escalate to critical importance. While you don't necessarily want to encourage failure, it is better when it happens early in the development and risk assessment process, allowing you to implement solutions as soon as possible and avoid a full-blown crisis.

Risk management is crucial, regardless of what type of project management you may choose — be it an agile, a waterfall, or a hybrid one. Don't rush through the process, but don't dwell on it too long, either. Most importantly, involve the entire team in the risk identification and classification process specifically, as this will help you get a bird's-eye view and a proper understanding of all the issues you might have to face along the way.

Chapter 6: Scaling Agile Projects

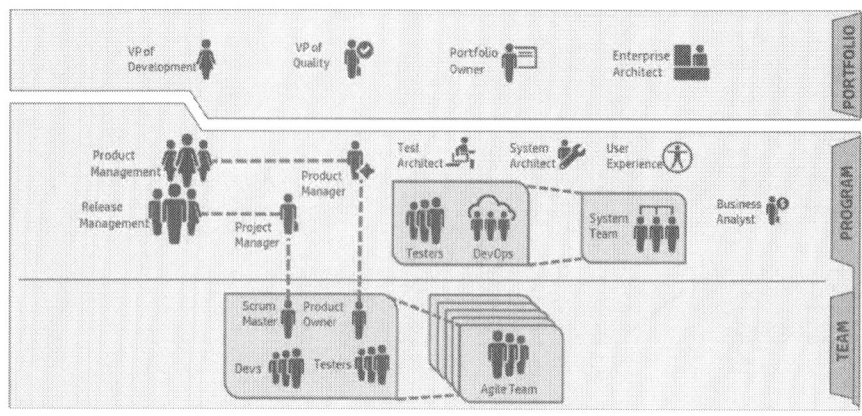

By this point in our book, we hope we have given you a comprehensive and compelling view of what agile project management is and how it works. Although we did not get into too much detail (mostly because we do not want to clog your memory with *too much* of everything), we do hope that you have grasped the main tenets of agile and how they work.

Most of the examples and the tips we have offered you so far function on all types of agile project management frameworks, and on projects of different sizes, too. Having gone through the basic concepts and ideas we have relayed so far, you most likely have a pretty good understanding of why agile works and how to implement it.

Scaling agile projects at different levels of the organization can be a challenge, but not an insurmountable one – especially these days, when even monumental organizations have already gone agile and showed that, *yes*, this is not a project management approach solely designed for those small, hip software houses.

Agile is present in every industry, in companies so diverse in size and scope that it is impossible not to believe in the power of this project management approach and all the benefits it offers. Sure, not all businesses use a pure agile approach — and that's okay. But the very fact that agile concepts have spread throughout the initially small space of software development companies and teams shows that there is a universal need for flexibility.

More than that, it shows that there is a universal need for change — in the way we work and in the way we perceive business.

This is meant to give you courage as you walk down the path of agile and agile scaling. It might not be easy. But it has been done before (and done well!), so you can definitely do it, too.

We will dedicate the last chapter of our book to this specific topic only: how to scale agile projects and go beyond what *you* have learned to pass it along to your team, your office, and, why not — your entire organization.

This chapter will discuss the challenges you might meet on your way to a fully *agile-ized* business, the types of agile scaling models you might want to consider, how to build agile teams that go beyond the basic principles and transpire into work ethics and ideology, the best practices for scaling up, and, ultimately, how to scale *out* when things don't go as planned.

We genuinely hope this chapter will give you a better insight into the more advanced world of agile project management and that it will fuel your thirst for more knowledge on how agile works at its most granular levels!

The Scaling Challenge

Change is always a challenge. Sometimes, even changing your outfit from day to evening can be a challenge — not to mention changing the MO of a company and its most intrinsic culture paradigms.

Scaling agile projects tends to be challenging from a series of points of view — most of which are related to the disruptive nature of agile, in terms of how it deals with change itself and how it embraces a new level of flexibility (one that, if we have to be honest, is quite absent in traditional project management).

We believe it is very important for you to be aware of the challenges agile scaling brings. You might not be able to remove these challenges upfront, but you can at least plan for them and treat them as risks, rather than allow yourself to be shocked and staggered when you have to face them.

Some of the implementation-related challenges you might encounter include the following:

- Lack of experience in the agile world. Reading and training yourself can go a long way, of course, and everything will get better as you accumulate more experience in the implementation of various types of agile frameworks and/or methods.

- Encouraging everyone to collaborate and coordinate. This can become especially difficult with distributed teams when they have to coordinate with each other in the development of a product.
- Understanding how sprints work and how not to lose the bigger picture.
- Getting the higher management in the organization to commit.
- Getting the entire organization to commit and align with agile principles.
- Setting up unrealistic expectations when it comes to both the implementation and the actual delivery speed.

Organizational culture lies at the very foundation of all these challenges. If you manage to change that, you can start your agile endeavors on the right foot and create the kind of agile project management experience that sticks to the core Principles *and* adapts to the specificities of your organization.

It is very important to mention that scaling an agile project is not always about the size of the project only. In fact, the very term "scaling" can refer to a multitude of situations.

Let's take a closer look at the three main types of scaling you might have to do and some of their specific traits:

Scaling at Team Level

Your team is the very backbone of your agile framework. You can read all the books in the world and you can create the most comprehensive training into the art of agile. But if your team is not on board with you, it will all mean you have just wasted a lot of precious time.

Scaling agile at team level should start small. It should start with your team before it then grows to other teams, as well.

The good news is that, although this might pose a challenge, a lot of companies and teams have already implemented agile practices in their workflows, even if they are not completely agile. For instance, you might see teams that focus on improving legacy code that are very traditional in terms of management, but use a Jira or Trello board to manage their tasks. That is, in itself, a grand step.

When you scale agile at team level, it is of the utmost importance to make sure that you are not pushing anything. Agile must feel organic — otherwise, it will put people in defensive mode and they will lose the jumpstart on agile project management *by the book*.

Show people the kinds of benefits agile can bring into their lives and take it step by step. Implement a Kanban or a Scrum board first, then the daily meetings, and slowly start to help people adhere to the values of agile.

Scaling at Project Level

Once you have scaled agile at the team level, it is important to move further and start implementing agile methods in the way you manage the project.

This includes the way you process product requirements as you receive them from the client or the product manager, how you split them into different stories, how you determine each story's specific level of importance, how you assign them to the various team members, and so on.

Scaling agile at project level is frequently easier than scaling at team level — mostly because once the team is on board, it is more likely that they will adhere to the new rules.

Keep in mind that different agile frameworks have different approaches when it comes to how projects should be managed. They are all very similar, but there are also differences you should take note of. You want to incorporate the methodologies that suit your specific conditions, but you also want to make sure you do this systematically, so that you can spot what works and what doesn't.

In case none of the pre-established agile frameworks are suitable for your particular needs, you will want to create a hybrid framework. And in order for that to happen, it would be great if you had a minimum documentation on what worked and what didn't (and why). It doesn't have to be a lot — just a few notes to remind yourself what to do with your next attempt.

Scaling at Company Level

Scaling at company level can be very easy or very difficult, depending on how familiar various teams within the organization are with agile methods and practices.

If the teams have been familiarized with agile techniques (such as the use of certain tools, for example), scaling agile across multiple teams or the entire organization will be a lot easier because people will be less skeptical — both when it comes to the teams and when it comes to higher management.

Scaling at company level means bringing agile to different teams — concomitantly or consecutively, as you see fit (depending on how familiar the teams are with the agile concept and the specific agile practices).

There is no simple recipe on how you can scale agile at team level, project level, or company level. Everyone does it differently, especially since every company and every project will be different in its own way.

There are a few characteristics that are common to all agile scaling projects, though:

- They usually start with Scrum and Kanban, since these offer a simplified version of agile and they are fairly easy to understand.
- They employ practices borrowed from other agile frameworks as well, including Lean project management, for example.
- They align their values and principles to those of agile project management.
- They tackle the main concerns related to security, performance, and dependency across different projects and teams.
- They aim to synchronize the delivery cadence of different teams to create a unified release.

Again, though — each company is different, each team is different, and each project is different, as well. You should always use your better judgment and employ those agile practices that fit the scope of your implementation and the specific characteristics of your organization and team.

Agile Scaling Models

By this point, you should not be surprised by the fact that there are specific models and methodologies you can employ when it comes to scaling agile, as well.

Although there are many agile scaling models out there, we will only introduce you to some of the most popular ones. The purpose of this section is to show you that while agile might be perceived as chaotic when you look at it from afar, it actually does follow its own internal procedures — including when it comes to agile scaling.

Some of the most commonly-used agile scaling models include the following:

SAFe (Scaled Agile Framework)

This is one of the single most popular agile scaling models out there - and, same as agile per se, it involves an iterative and incremental approach. When SAFe is used, there is a focus on coordinated processes and on being both knowledge-based and interactive.

SAFe is used when you want to leverage agility and lean development, as well as when you want to coordinate the agile scaling process. The main purpose of this model is to align the different team activities to the organizational needs and objectives.

SAFe can be applied to any specific agile framework because it aims to define the general agile principles and implement them, rather than specific practices that might pertain to one or two agile frameworks only.

In general, it is important to note that complex instructions can make SAFe less compatible with agile methods and as such, you might want to look into other agile scaling methods instead.

DaD (Disciplined Agile Delivery)

There are quite a lot of similarities between SAFe and DaD (Disciplined Agile Delivery). However, the latter tends to be more "hybrid," rather than purely agile. Its main goals are the goals themselves (as *meta* as that might sound). DaD is more enterprise-friendly in the sense that it focuses on making decisions based on incremental and iterative processes, rather than merely making definitive, prescriptive decisions based on spreadsheets and assumptions only.

When you're scaling agile, DaD will most frequently be used when it comes to determining the basics behind your project: the schedule, the tools, the IT platforms you will use, and how often (and how!) you will maintain the system.

LeSS (Large Scale Scrum)

Large Scale Scrum (LeSS) is one of those agile scaling methods that is more inclined toward the agile project management methods (unlike SAFe and DaD, which pertain to a more general/traditional project management approach).

LeSS is used in large-scale Scrum development. If there are fewer than ten teams, the main roles in Scrum will remain the same (the Scrum Master will play the same role, the product manager will play the same role, and so on).

Furthermore, in LeSS, team meetings are replicated across the different teams either by using means of online communication or by offering everyone retrospectives from the point of view of one team.

If the project involves more than ten teams, you can use a version of the LeSS that has adopted extra coordination functions. In these cases, the traditional Scrum roles will be split according to the areas the teams are located in:

- The Product Owner becomes an Area Product Owner
- There will be an Overall Sprint Review to include each Area Product
- There will be an Overall Sprint Retrospective that includes the entire program

Nexus

Similar to LeSS, Nexus was created specifically for agile scaling purposes. Also like LeSS, Nexus is set upon the Scrum model and allows the different teams of a scaled agile project to coordinate their dependencies and integrate their results. Furthermore, Nexus also comes with plan, launch, and software management elements, as well.

The main purpose of Nexus is that of incremental integration into the coordinated product. Nexus is very much like LeSS in many respects, but it works on a smaller framework — it can only support up to nine teams.

LeadingAgile

LeadingAgile is an agile scaling model focusing on scaling agile across the various departments of an enterprise. Rather than offering a specific blueprint for companies to use when they integrate agile across multiple departments, LeadingAgile focuses on providing consultancy services to companies who want to do this.

This ensures that the scaling model will be completely personalized to the organization's individual needs and goals.

The Scrum of Scrums

It's not that the other agile scaling methods are not popular — it's just that for most people (and even more so for beginners), the Scrum of Scrum approach makes the most sense.

The Scrum of Scrums does exactly what its name suggests: it brings all Scrum Masters together and replicates the practices of a small Scrum team across multiple Scrum teams.

One team member (the Scrum of Scrums Master) is delegated to represent the team when the Scrum of Scrum meeting takes place. This might be the actual team Scrum Master or any other person in the team. Their job is to coordinate with the other delegates from the other teams to unify their processes and create an agile framework at *scale*.

Building Agile Teams

Beyond all scaling issues, the *team* is the very kernel of how scaling happens. All the theories in the world mean absolutely nothing if you cannot build an agile team you can scale from.

Building agile teams starts with you and the kind of example you set. As mentioned before, it is of the utmost importance to create the kind of environment that helps people flourish in their adoption of agile project management and all the techniques that come with it.

We will dedicate this section to help you understand how to build the right agile team. Same as with everything agile (or everything business-related, for that matter), building an agile team cannot follow a specific recipe.

What you can do, however, is follow some basic tips on how to build an agile team that actually *works* (in the productivity sense and in the sense of the smooth flowing of the entire team, as well).

Be Patient

Patience is a real virtue, especially when it comes to changes as large and as daring as agile deployment and scaling.

You cannot expect your team to absorb all the agile practices in one week, and you cannot expect their peak performance levels to be reached in a matter of days. Not everyone will be fully on board from the very beginning, either — some may perceive agile as a waste of time, others might be worried that it will lead to micromanagement.

Be patient and answer everyone's questions as they come. It's normal for your team to be doubtful — human beings have been resilient to change ever since they settled down in communities, and modern people make no exception from this rule. It is in our DNA.

In general, agile group forming tends to follow four stages:

- Forming — At this stage, everyone needs guidance and the roles of the people in the team are not yet fully formed.

- Storming — At this stage, people start to understand the inner workings of agile and how decisions are made. However, the relationships between the various roles are still quite unclear.
- Norming — At this stage, processes start to become more optimized and people understand the relationships between them.
- Performing — At this stage, the team starts to be genuinely efficient. Each member knows what they have to do and how to do it, and they understand the mechanism behind their manager's job.

You can expect your team to fully grasp the magnitude of agile and its benefits somewhere along the process. But you cannot expect them to perform at peak efficiency before they reach the Performing stage.

Be patient with this. Don't rush and don't push — it might actually have the exact opposite effect of what you're aiming for!

Adapt to Change

Change is why agile took birth, so it makes all the sense in the world that, as the project manager guiding your team towards agile, you should be adaptable as well.

You might set out to implement a pure Scrum methodology in your team and realize that it simply doesn't work — or at least, not yet. Be flexible and re-adapt the methods and practices to suit your particular case. Of course, you should still stay true to the Agile Principles —

fortunately, they too are flexible enough to leave room for personalized interpretations.

Show your team that you are not afraid of change, and that you are ready to embrace it no matter where it comes from and how it occurs. If you do this, they will follow your lead. If you panic, however, or if you stick to a very prescriptive agile approach, you might find that your team doesn't like it.

And the next thing that will happen?

They will leave the team, they will constantly try to sabotage your agile implementation efforts, and they will go the exact opposite way in terms of honesty, self-discipline, and self-organization.

That is where the so-called agile "chaos" occurs!

Be Results-Driven

Good agile project managers and teams don't focus on mistakes — or on who did them, or on how many times they were done.

Instead, they focus on results.

When each of the team members has a pretty clear goal in mind (which you will set with them), they start to shift perspective from *he-said/she-said* to taking responsibility for their own actions, implementing the lessons, and ensuring that they will not repeat the same mistakes.

Never try to place the blame on anyone. Find the source of the problem and make sure you eliminate it in the future. The end result (customer satisfaction!) matters more than who made a mistake and when!

Don't Miss the Bigger Picture

When you work at such granular levels, and when each product feature is split into multiple tasks, it is very easy to miss out on the bigger picture.

That is precisely why you should have a clear product vision in mind (and on paper).

Both you and your team members should make sure you constantly aim to achieve the *final product*, rather than get lost in the woods and led astray.

Constantly reminding yourself and your people that the bigger picture matters more means that you will all have a stronger sense of belonging in the team. As a consequence, each of the team members will be able to find intrinsic and extrinsic motivation to stick to agile and everything it brings along (including disarming honesty).

Be Transparent

As mentioned at the beginning of this book, agile doesn't sweep problems under the carpet — precisely because it understands that doing so can lead to a more serious issue in the future.

Transparency is a basic principle in agile project management. Everyone knows what everyone is doing, and the Daily Scrum plays a huge role in this. Plus, everyone can collaborate with everyone and can constantly ask questions about the progress of certain tasks, so that they can coordinate properly.

Results are transparently relaid to the team, as well. There is no point in lying about bad results or good results — the only point of showing these results is to improve. Because, yes, agile project management is all about continuous growth and iterative improvements, right?

Ask for Feedback

Asking for feedback might feel odd to some, but it is something that will help your team understand the human value of agile project management.

In a nutshell, asking for feedback will:

- make you more trustworthy
- help you grow
- encourage team members to ask for feedback, as well
- promote a culture of open communication
- promote a culture of constant growth

By asking for feedback, you set the example. You become the ideal agile team member and set an expectation in the way your team members mirror your behavior.

Oh, and by the way — when you do receive feedback, make sure to implement it, no matter what it may be related to. It can help everyone: you, your team, and your organization, too!

Give Feedback on a Regular Basis

Feedback loops are quintessential in agile project management — but not solely when they come from the customer.

Obviously, you want your customer's feedback on each iteration's release, so you can improve the product and send in a better version every single time.

Establish regular feedback sessions with your team. Run group feedback sessions and one-on-one feedback sessions, as well. Both have their own purpose and they both can help you grow a healthier, more productive team.

Show Trust — Always

You can't expect people to trust you and what might seem like "agile mumbo-jumbo" if you do not display the same behavior you would want to see in them.

Trust is important in every type of agile team under the sun. It helps people collaborate more efficiently, it helps people communicate better, and it helps teams deliver faster, better, and more in tune with both the budget and the client's expectations.

It might not be easy avoid looking over your developers' shoulders to see if that typing is all about hard codes or simply long Slack chats with the girls in Accounting.

But you have to do it. You have to show trust and only start asking questions when that trust has been broken.

Believe it or not, people are not inherently lazy or unmotivated. Positivity tends to return the same level of positivity.

Have Clearly Defined Roles

It is very important to note that many people believe *agile* means that team roles are not clear and that the project manager will end up coding, while the developer will end up running the office manager's tasks.

While that might be somewhat true in very young companies that cannot afford to pay so many salaries, it is definitely not a mark of *agile*, but a mark of working in a start-up.

Make sure your team roles are clearly defined and that everyone knows what everyone else is doing:

- The Product Owner/Product Manager intermediates between the customer and the Project Manager and their team
- The Project Manager handles delivery times, budgets, task assignation, feedback, and reporting
- The Developer writes code
- The QA Engineer checks the code

... and so on.

Maintaining clear roles will help with productivity in the long run, because it will allow each team member to focus on what they do best: communication, management, or development.

Encourage Team Members to Help Each Other

This is not to say people are not allowed to help each other when needed, but this should be something they do *aside* from their actual tasks because they actually *want* it, rather than because they feel obliged to do it.

Collaboration is not always about everyone working in their own cubicle and not minding each other's business unless they are somewhat forced to. Collaboration means sharing ideas, knowledge, and helping each other.

What you should pay attention to is that everyone does their job *first* and then helps other team members. Doing otherwise is not productive and it can lead to a massive waste of time.

Make Your Office a Comfortable Place

We have already discussed this, but we will mention it here as well. The office is where people spend one third of their working days. It needs to be the kind of place they actually want to come to — a place that fosters productivity, and even more than that, a place that fosters comfort and even friendship.

Open-space offices have grown in popularity precisely because they allow people to communicate better — and they are a fundamental trait of agile teams, especially in those companies that can afford investing in a space that is very agile in nature and encourages everyone to work efficiently.

As we mentioned, there is a very good reason so many software development companies invest in game rooms and relaxation rooms, too: they work. You might feel that giving a bunch of developers a game console will irreversibly lead to them wasting atrocious amounts of time trying to beat each other's FIFA scores.

However, in reality, people are motivated by the short breaks. They tend to get their work done faster and better. They tend to be happier at work. They tend to collaborate better with each other. And, ultimately, they tend to have more trust in you and the entire organization (which also means they will be less likely to leave the company, as well).

Maintain the Stability of the Team

In an ideal world, you would work with the same team for extended periods of time. The more time you spend with your team members, the more you get to know them and the more you get to understand what makes each of them tick.

In the real world, job mobility is an actual issue — particularly in the software development world, where it seems that there are more open positions than people trained to fill those jobs.

The stability of your team can influence the success of your project and there is no way around it. When people keep coming and going in and out of the team, you don't get to actually build that *agile* mindset with them. You don't have a uniformly developed team, where each member performs at their peak level of productivity and efficiency.

You should do everything in your power to keep team members, especially seniors who are hard to come by.

Listen

Trust, collaboration, and proper communication cannot happen in a world where people don't listen to each other.

It is, thus, of the utmost importance to make sure that you genuinely listen to the people in your team. Listen to their questions (and answer them), listen to their worries, listen to their skepticism, listen to their feedback and their praises.

Listen. Listen more than you talk. That is how you genuinely get to know your team and create an environment where everyone is encouraged to be honest with themselves and with everyone else!

Believe

You cannot set a true agile example if you are skeptical of agile in general. If this change is imposed to you, then you should try to read as much as you can about the benefits of agile and how it can help you find more satisfaction from your work and create a better work-life balance for yourself and for your team members.

If you want to breathe the agile spirit into your team, you have to believe in it. Even a slight trace of mistrust in the method can be easily *felt* by your team. And that means they will become more skeptical themselves, and more likely to be dishonest and take advantage of you and the other team members.

Believe in agile and its power to transform you, your team, and your organization. There are so many examples out there, it is hard to ignore the functionality of agile!

Of course, there are many other tactics you can use to make sure your team becomes more and more agile with each iteration. At the end of the day, it's all about making sure each member understands what they have to do, and providing them with the right space and mindset to achieve that.

Everyone moves differently. People are stimulated by different motivators. As such, there cannot be an actual prescription on how to build agile teams. Move along with the beat of your own team and adapt — because if you aren't *flexible* as an agile project manager, you lose the *agile* in your work. And when that happens, it can all go south very quickly and very dramatically.

Agile teams don't happen overnight. They are made up of the people who actually want to be there. They are consisted of the energies that flow between the different team members and how they interact.

More than anything, agile teams happen only when their leaders live up to the agile philosophy. You cannot expect people to be honest with you if you don't pay them in the same coin. You cannot expect people to trust you if you don't trust them. And you cannot expect people to take responsibility for their own actions if you don't do it, either.

Agile teams are born when agile project managers master the art of being flexible and trustworthy. They are born when their PM's approach goes far beyond methodologies and strategies. They are born when they start to believe in agile by the power of their own project manager's example.

Scaling Up: Agile Practices

In addition to the specific methods we have discussed throughout this chapter, and in addition to the ways in which you can make sure the team you scale up *from* is truly agile, it is also important for you to understand some basic agile practices for scaling up.

Knowing (and applying) these practices will help you gain more control over your scale-up efforts — and as such, it will improve your odds of success.

Same as in the case of building an agile team, nobody can offer you a strict, prescriptive recipe on how to do it. What we can do, however, is present you with the best practices used by people in the field. Along with everything you have learned thus far, they should help you make sure your scaling up is successful.

Do keep in mind that these practices cannot ensure absolute, infallible success for your agile endeavors. They can, however, establish a healthy environment for your agile deployment and scale up.

Start with the MVP — The Minimum Viable Product

One of the best ways to make sure you have something good to scale up from is by starting with the Minimum Viable Product.

For those of you used to traditional project management, this might not make much sense — in waterfall, you deliver the entirety of the product at the end of the development cycle, once it has all been tested.

In agile, however, delivering an MVP means that you will have something to grow from. Not only will you be able to show your customers something they can actually lay hands on (or, better said, lay mouse cursors on), but you will also give your team a boost in morale, and you will be able to draw the right feedback and the right conclusions to ensure better performance as you continue to develop the product.

When the customer has an MVP in hand, they should be informed of the product's current status. And they should be encouraged to give feedback on whatever they might want, so that you can include their input in the subsequent development iterations.

Scaling up from an MVP tends to be easier, because you will only take the product requirements and go more in-depth with them on a foundation that is already built (and already analyzed by the customer).

Use One Product Backlog

If you want to scale agile across multiple teams, it is important to have only one product backlog, even when these teams are very dispersed (geographically and in terms of the kind of work they do).

Having one backlog helps team members trust each other even when they are worlds apart, and it helps you see the bigger picture (instead of trying to puzzle it back together from smaller backlogs).

Plus, let's face it: it is downright easier to only use one product backlog from which all tasks spring out. Why would you even bother handling two, three, or 20 different backlogs?

If you do decide to only use one product backlog, it is quite important to mark every task accordingly. In Jira, for example, you can mark each task according to the department or compartment it pertains to. This might not seem like much, but it can definitely help avoid nasty confusions and misunderstandings.

Make Sure Everyone Understands the Process

You can't scale across multiple teams if they don't understand the processes behind agile and the processes behind the actual scale-up.

Take your time to show those involved how it all works and how it all looks, how you plan on bringing together all the features, and how you want them to fit in the entire picture.

Make sure everyone understands what each meeting means (including the daily meeting and the Scrum of Scrums, if you choose to have these). Make sure everyone understands the importance of being honest about their tasks and the time it takes to lead them to an end. Make sure everyone is of the same mindset in terms of why agile is important, how it helps, and the kind of mentality they should adopt if they, too, want to reap the benefits of agile project management.

This is not to say all of your teams should be brainwashed into the same kind of thinking — this is just to say that every single team member in every single team, no matter how dispersed, should aim for the same goals as the "core team" back home.

Emphasis on Collaboration

We have said this a thousand times, but we cannot emphasize it enough. Without a spirit of collaboration, there is no agile — not in a small team and not in a team that is scaled up.

No matter where on Earth your teams might be located and no matter how small or large your company may be, collaboration is absolutely crucial because it will help you create a product that is uniform and functional across all verticals.

Trainings, Courses, and Certifications

Trainings, courses, and certifications help with scale-up projects because they help people understand agile at a higher, more in-depth level.

At the same time, making sure that all teams in all locations follow the same trainings and courses and that they get the same certifications can also be helpful when it comes to building a uniform culture of agile project management.

Yes, agile will be taught the same everywhere on Earth, because its rules are universal. But when all of your teams can reference back to the same materials, it is far easier for them to collaborate, communicate, and, in general, adopt agile project management not as a *must,* but as a way they can grow, too.

Improve Your Development Infrastructure

Scaling up agile projects also means that your technical development infrastructure should be uniform across all teams, regardless of whether or not they are located in the same place.

Your development infrastructure should be able to support the incremental and iterative nature of your development process. For example, you might want to consider a microservice architecture for your development infrastructure, as this will allow multiple teams to deploy and release their increments without being codependent.

Once all the "bits" of the product are deployed, they can be brought together and delivered to the client (as a Minimum Viable Product, as a story, or simply as a reinforced and improved version of a feature the customer has seen before).

Stick to Smaller Teams

For some of you, it might make sense to bring multiple smaller teams together in a larger group. However, scaling up is not about numbers, but about proportionality.

In other words, your scale-up strategy should maintain the same small number of team members because they will collaborate easier this way, and they will be able to mirror the iterative process in a better, more efficient way.

Just think of agile in general and why it has chosen to promote working on very small, bite-sized chunks of projects: they are easier to develop and they are easier to adapt, should anything have to change.

Apply the same rule to the size of your team, as well. The smaller they are, the more the Scrum Master can handle them and the more granular they can get with the planning and the task assignation.

Coordinate the Production Time with the Iteration Length

If your customer is aware of your agile approach, they may anticipate that the delivery might not be made at a very specific time, but they will know it will be done before the deadline.

It is quite important to make sure your product development timing is well coordinated with each iteration. In general, an iteration (or Sprint, if you are using Scrum project management) does not take more than a couple of weeks. By the end of each iteration, your team should have entirely finished a chunk of product.

The production time and the iteration length are always to be determined together with the customer and the team. It is essential to make sure you are realistic and sustainable about this from the very beginning, as this will allow you to plan accurately and keep both the customer and the team happier.

Make Sure Someone Is the Product Owner

The Product Owner/Product Manager role exists in agile project management for a very good reason: you actually need it.

In smaller teams and start-ups, the Product Owner might also be the Project Manager. We understand why this might be the case, especially in companies that cannot afford to hire a separate Product Owner. However, you should understand that a PO can really expand the relationship between your business and the customer, and that they can bring real value.

The Product Owner should be the liaison between your company and the customer. This person will be the one handling the communication between the company representatives and the actual client — and this means they are also the ones who will "translate" the client's needs into actual product requirements the PM can take to their team.

Everyone has a very clearly established role in agile project management, and the Product Owner makes no exception from this rule.

Synchronize the Iterations Across Teams

If you want your agile project to be perfectly scaled up, it is quite important to make sure everything is uniform — specifically those parts that are essential in terms of customer satisfaction.

As such, you should also make sure that your iterations/sprints are synchronized across all the teams involved in the project. This will allow everyone to coordinate and collaborate at a higher level of efficiency.

Use the Right Tools

There is a good reason we dedicated an entire chapter to exploring some of the most popular agile project management tools out there: you need them.

Sure, you can do agile with a spreadsheet, too, but it will waste you and your team a lot of time, and it will make inconsistencies across different teams much more recurrent and poignant.

The right tool can be a life saver in agile project management. What is important is to make sure that all of the teams you have scaled up to use the same tool. For instance, if you use Jira, everyone else should use Jira, too — this will allow everyone to see the same backlog, to track the progress of each task, and to take responsibility for their own actions in front of all the unified teams.

Face to Face Meetings and Team Buildings

We live in a highly technologized world that allows us to communicate across distances in a matter of seconds.

Just look at the book you are holding. In itself, this book is an act of communication — one you downloaded in a matter of seconds and you can read regardless of where in the world you may be.

It's pretty amazing, and it has definitely pushed the boundaries of tech development further and further, as companies now have access to a larger pool of talent than ever before.

At the same time, it is crucial to understand that, even with all these tools and even when you are connected to the fastest internet in the world, you can never replace face-to-face meetings.

Arguably, the basics of agile project management were laid out in an era where video calls were very low-quality and nearly impossible in some parts of the world due to poor internet connection. This has changed, and as such, the relationship between dispersed teams have improved, as well.

Even so, having regular meetings and team buildings can help a lot — precisely because it will make for better relationships between the teams and the members that comprise them.

It is an extra-cost for your company, for sure.

But it can pay off so well!

Scaling Out: Disturbed Projects

If you have reached this point in the book, congratulations! We truly hope you have learned a lot about the nature of agile projects and how they are scaled up across the different teams and, ultimately, across dispersed organizations.

As you learned at the beginning of this book, agile project management can bring a lot of benefits — both to you as a PM and to your team and organization.

There is a myriad of further advantages that derive from everything that was explained throughout this book. Yet, we would be terribly unfair towards you, as our reader, if we didn't point out that, well, agile doesn't always work.

There is no point in lying about it. Agile is not always efficient — but not because agile is flawed in itself. Most of the time, the reasons that make agile *not work* are connected to extrinsic contexts and circumstances.

Some examples of the kinds of situations that might prevent agile from working at its peak efficiency include the following:

1. There is no or little support from higher management. This is one of the most common reasons for agile failure — and one of the saddest ones, as well. On the one hand, it is quite easy to understand why higher management might not have time to jump the agile train — they are already handling many other things. On the other hand, however, agile project management can make their lives easier down the road, too.

   When you lose higher management support, trying to transform your organization to an agile one can feel like the work of Sisyphus. And it is precisely at this time when you might want to consider a hybrid approach of some sort — one that doesn't involve higher management as much, but still stays true to the basic agile Principles.

2. Lack of clarity. This pitfall is especially true in larger organizations, where information can be easily lost between the emittents and the recipients of the communication channel. Lack of clarity can lead to misunderstandings, completely ruined sprints, and, eventually, it can delay the entire project, lower the quality of the deliverables, and push the budget over the limit.

3. Sticking to legacy methods. It is easy to understand why you might want to stick to legacy methods that maintain a bridge between the traditional waterfall approach you have worked on so far and the new one you are trying to adopt. However, when these legacy methods go against the basic Principles of agile (including the ruling one, *flexibility*), it is time to question them. If you don't do this in due time, you might label agile as useless for your company — and as such, you might drop your endeavor of implementing it.

4. Poor understanding of agile. We strongly encourage you to read as much as you can and to attend as many courses, seminars, and trainings as possible. Agile cannot be learned in one sitting — it is a process you have to fully embrace. Our advice is to not jump into actual agile deployment before you fully understand all its concepts, from the ground up. The same goes for your team members and for all the other stakeholders in the organization, as well!

5. Lack of Product Ownership. As mentioned in our previous section, having a Product Owner is valuable. Yes, there are agile companies that function quite well without a dedicated Product Owner. But the lack of someone to hold this role can also be your demise, as it will charge the PM with too many tasks and render him or her unable to run those that are in his or her actual job description.

6. Insufficient testing strategy. Development is one thing, but since agile is very focused on continuous improvement, a consistent and comprehensive testing strategy is absolutely crucial. Do invest in people who can test each iteration down to

its smallest details, and invest in a strategy that will allow you to deliver working software that is as close to the customer requirements as humanly possible, with every iteration. Otherwise, you might find yourself trying to constantly repair previous iterations in an endless cycle.

7. Large company size. It's not that agile cannot work in large organizations. It does. You know it does, because mammoth companies have implemented it successfully (Microsoft and Oracle are just two examples). However, when a large enterprise is too entrenched in traditional ways and when higher management is not actively involved in adopting the change, agile will not work. It is a sad truth, but it's one we all have to accept.

8. Too much flexibility. Too much of anything can be harmful — and flexibility makes no exception. It's one thing to be flexible, but it's a completely different thing to be downright laissez-faire. It's one thing to adapt, but it's a completely different thing to never plan out anything and never stick to a plan. Too much flexibility can kill agile, because it will make it inefficient on all levels.

Of course, the specific situations in which agile can fail vary from one company to another. These are some of the most common issues that lead to failed agile deployments — and it is more than worth being aware of them so you can avoid them as much as possible.

Most of these situations are, indeed, avoidable. But if there is one thing you cannot change about your organization, that is its size. Indeed, scaling up in very large enterprises can be truly challenging even when you employ all the best practices in the field.

So, what happens when agile scaling up seems to have failed for good?

Simply put, you back out and rethink your strategy. In other words, you scale out.

If scaling up happens vertically, scaling out happens horizontally. To put it in very simple terms, "scaling out" in agile means "adding more nodes" to your system.

After reading this entire book, this might seem like the least intuitive thing to do, but it has been proven to work. When you add more nodes to your entire agile system, you allow everything to be more granular again, while sticking to the basic agile Principles.

Indeed, more nodes in the management system might be more expensive, and it might make bottlenecks a more intricate game to play (which goes against the concept of simplicity all agile projects should follow).

However, it is a compromise worth making if it means that you will be able to maintain agility within each team and allow the different teams to continue collaborating according to the basic agile Principles.

You shouldn't give up. Implementing agile in a team can be difficult — implementing at the level of an entire enterprise can feel downright crazy, especially when that organization is fully embedded in the traditional paradigm.

It's not impossible, and it can be done. The key lies in finding the right agile framework (be it a purebred agile or a hybrid). More than anything, though, the key lies in constantly improving yourself as a project manager and as an agile promoter.

Studies show that, in 2018, agile-managed projects were 28% more successful than waterfall-managed projects (PWC, 2019). It might not seem like much, but when you work in an environment that is ever-changing and when you truly want to push yourself and your team beyond the limits of a traditional paradigm, that number can mean everything.

Are you ready to make the agile leap, too?

Conclusion

By this point, you should know what agile project management is and, well, what it isn't.

Agile project management is a mindset more than a specific method. It is a way to look at work and life. It is a transformation and a continuous process of improvement at the same time.

Agile project management is not a one-size-fits-all kind of solution. Then again, nothing is — and if you ever come across someone who promises to hold the solution for *everyone*, run like the house is on fire. There is no such thing as a universal recipe for everything and everyone.

At the beginning of this book, we aimed to explain the main concepts behind agile project management. Hopefully, we've managed to do this. But even more than that, we hope we've managed to open your eyes to a world of opportunity where things aren't always black and white, agile or waterfall, successful or unsuccessful.

Agile project management's main lesson is flexibility. It lies right there, at the core of the 12 Principles, at the heart of the Manifesto, and at the very foundation of every single agile framework under the sun, from Scrum to XP and back again.

Even if you close this book and decide to never try agile project management in its purest form, we still hope you will have learned an important lesson on the importance of being adaptable in the way you work and the way you deliver your product.

It is perfectly understandable why you might decide purebred agile is not for you and your company. We do not try to sell agile here — all we are trying to do is help people see things differently, from the perspective of continuous improvement, rather than a perspective that is clenching its jaws and frowning upon everything that is not pre-planned, pre-established, or predestined.

Agile doesn't work for everyone, and it's a truth we have learned to accept. We are genuinely big fans of agile project management and everything it stands for, sure — but we also understand that companies, people, and teams function differently. And that's okay.

Before you toss the entire idea of agile project management to the trash bin and close this book, though, we want you to take a moment and think of the myriad ways agile could work for you.

Think of how great it would feel to come to work every day to a team that does everything out of sheer passion and dedication. A team that does everything because they actually want to do it, rather than because you are *watching* them.

Imagine how it feels to continuously deliver products that make customers happy. Products that help businesses move forward. Products that make end-users live a better, easier, happier life at their own jobs.

Imagine how it feels to be always useful, rather than a pestilence constantly breathing down your developers' backs.

Imagine how it feels to do your own job knowing fully well that you will make an actual impact down the line, and that you aren't perceived as just another email pusher.

It does feel nice to imagine all these things, right?

Well, then, maybe you should reconsider agile project management once more. It may not work in its purest form in all situations, but that doesn't mean you cannot "borrow" concepts from agile and apply them to a more traditional approach.

The reason we focused so much on agile scaling is because we genuinely believe it can be done. As we mentioned in one of our examples in the book, Microsoft and Oracle did it — and this proves that, with the right approach (and, arguably, plenty of support from higher management), you can actually make it, too, no matter how large or small your organization may be.

The key is, as we have also mentioned, believing.

The right approach is out there, waiting for you to discover and rediscover it, waiting for you to adapt it to your own specific needs, and to the ways in which your organization has been doing things until now.

Giving up now means you are not giving yourself, your team, and your organization a chance to do things better, more flexibly, with a bigger impact down the line.

We genuinely hope we convinced you to give agile project management a chance. It may not be easy to make the switch from waterfall to agile (or pick up agile as your first-ever project management approach, for that matter). However, determination and passion can go a very long way.

In the end, determination and passion are why agile was born in a decade that needed it more than ever. And, over the course of the decades since, agile has proved, time and again, that its core Principles can actually transform the business space.

People are happier at work these days than they used to be. They live their 9 to 5 in better environments. They are acknowledged as human beings, not robots working in cubicles. They are trusted. They are learning that, yes, you can be honest about your mistakes and that, yes, there are people who believe you can do better.

And all of this is due to agile project management. We might not be fully aware of the impact this approach has had on the lives of millions of people out there who woke up one day to go to do their job in a way that is actually humane and growth-focused.

We do know, however, that *numbers* show the impact of agile across different organizations, in different spaces.

And we do know that the very nature of agile project management allows you to adapt it to your own needs.

We invite you to learn more about the fascinating world of agile and how it can be implemented in a variety of situations. Our next book will focus on the most popular agile frameworks out there: Scrum, Kanban, XP, Crystal, FDD, DSDM.

These names might sound downright scary at first glance, but we plan on helping you understand how they work and how they can impact you and your organization.

At their very core, all of these frameworks follow the same basic Principles and the same Manifesto. They do display differences, though — and it is precisely in those differences that you might find your saving grace, the specific agile solutions you need to implement in your organization to help it move to the 21st century in terms of how it creates products and delivers them.

We definitely understand that agile project management might sound idealistic to many of you — particularly those who have been used to the strict structures of waterfall project management.

What we do want you to understand is that we are not selling a dream here. We are selling hard-coded facts: agile works. It may not always work as it is, and you might have to make adjustments to it. But, down the line, it can really reshape your entire perspective as a project manager, as a worker, as an entrepreneur, and, ultimately, as a customer of products that have been built using agile methodologies.

We started this book with a brief introduction into why agile was needed and how its birth took place. In a world that moves faster every day, agile is more needed than ever.

This goes beyond software development, and it stretches out to every single industry you can imagine. Schools can now better address their needs. Hospitals can manage their supplies more efficiently. And even clothing companies can help reduce waste to create a better world for tomorrow.

Agile impacts the world we live in — and it does so in a silent way, from behind the spreadsheets and the Jira tickets we all frown upon. Behind every large change you see in the world, there is an agile PM working their heart out to make sure products are released on time and according to the specific needs of all the stakeholders involved in the project.

Agile project management is not about developers who stand up for two minutes every day.

It is about the very nature of our modern world —ever-changing and in dire need of a new mindset. A mindset that focuses on the good in people, rather than their laziest and most inefficient inclinations. A mindset that focuses on trust and hope, on adaptability and acceptance, on acknowledging there is life after work and *at* work.

Agile has changed the world in more ways than most of us imagine. And it will continue to do so as more and more businesses start embracing it.

It is said that the millennial workforce is now ready to take over the market. Agile was there before millennials, and it prepared the field for a generation of people who refuse to work without meaning, to be enclosed within strict rules and regulations, to be hidden away in cubicles and away from anything that might stimulate their growth.

The seventeen people who lay the official foundation of agile project management had no idea what the new generation would do. And they likely had no idea they were changing the world, either.

All they wanted was to do things more efficiently, with less headache, and more profitability.

And as such, they wrote down the 12 Principles of Agile Project Management.

Nearly two decades later, the Principles are everywhere, even in companies that do not admit to going agile all the way. They are in the way your office is built and in the way your boss treats you better every day at work. They are in the very air you breathe and in the way your local hospital manages its operations for better and more efficient services.

These days, the agile Principles run through the blood of everyone on the work market. And, although there may be industries where agile will never work full-time, even the oldest and the most traditional companies are opening their eyes and their hearts to a world that is focused on *results*, rather than *documentation*.

Agile is bound to change your life, as well. At individual level, agile project management will make you a better person because it will show you just how much flexibility can help in *everything*.

At team level, agile project manager will help your team members genuinely grow — not just as developers, but as a human being, too. They will be more inclined to take responsibility, they will be happier, and, overall, they will be more satisfied with what they accomplish each day at work.

At company level, agile project management will boost productivity, profitability, and customer satisfaction.

There is really no way you can lose when you embrace agile, even if you can only do it in its hybrid form.

We hope the book at hand has helped you understand the importance of understanding agile at its very core. We hope you have lots of questions and that you will continue to read more about agile project management (and its specific frameworks).

But, more than anything, we hope we were useful and that this book was the catalyst of genuine change in your life.

It won't be easy, as we have reiterated throughout the entire course of this book. But there truly is value in adopting agile project management. The thousands of companies that are already using it are evidence of that. But it will be your own experience with project management that will ultimately prove it to you.

From the way you handle your own tasks to the way you plan out the projects you manage at work, agile will switch your mindset to one focused on results, transparency, and genuine adaptability.

And we hope this book is just the beginning of it all!

Happy learning onwards — there's a long, winding, and slightly confusing road ahead of you, but once you know why you are walking upon it, you will start to truly love its ups and downs, its steep curves, and its splendid landscapes.

Agile Project Management

Methodology. A Comprehensive Beginner's Guide to Scrum, Kanban, XP, Crystal, FDD, DSDM

By: Sam Ryan

Introduction

It's easy to dismiss a project manager's job and put it in a little stereotypical box painted in spreadsheet columns and rows that multiply to an endless infinity.

It's hard to actually understand the massive amount of work a project manager has to handle on a daily basis and the kind of pressure they have to face when they are torn between the team they spend eight hours every day with and a higher management level that is constantly asking for reports and profitability graphs.

If we have to be completely honest, project managers are modern-day heroes. They don't wear capes, and sometimes they can be downright annoying when they ask you about the status of your latest task. They don't have any other superpower aside from resilience and infinite patience. And they definitely don't have TV shows dedicated to them (although we would definitely back up a series based on the tumultuous life of an average PM in any software development company in the world).

Instead, project managers are frequently avoided and mistreated, misunderstood, and downright minimized. From the backbone of every company under the Sun (including *air flight companies,* mind you!) to spreadsheet carriers and email pushers, project managers seem to be in a constant limbo.

Truth be told, being a PM is difficult. Leaving aside the mad Excel skills you have to possess, being a PM is more about the people than it is about the reports—now more than ever, especially with agile project management taking over the world in a storm of variations.

The first installment of our Agile Project Management book was dedicated to exploring the fascinating world of agile and everything it has changed ever since its official inception in 2001.

Because we want you to have a solid foundation as you start reading the second installment of our Agile Project Management series, we will quickly move through the basic concepts we have already elaborated upon in the first book. If the volume at hand is where you stumbled upon us, we DO advise you to read the first one as well, as it will provide you with a treasure trove of information on where to start in the world of agile project management and how to tackle some of this framework's main challenges.

We started our first book with a quick history of agile—not because we like boring people in any way, but because we believe the history of agile project management is very tightly connected to the *needs* it came to solve. More specifically, agile project management was born on the back of a world that was speeding up like never before—a world that was completely incongruent with the strict limits of the traditional project management approaches practiced until then.

By the beginning of the 1990s, the first agile methods were slowly shaping up. Some of them had been used for decades already; others were born in an industry that was as new as the second half of the 20th century. In essence, they all came together under the Agile Manifesto at the beginning of 2001, and ever since then, they have continued to evolve and mix with each other (as you will see in the book at hand as well).

We also tackled the advantages of agile project management and why so many businesses are embracing it, either in its entirety or as part of a hybrid approach. From being able to deliver better quality to improving the quality of your team's life (at work and outside of it), agile project management has changed quite a lot. We are daring enough to say that agile has shaped up the world we currently live into a point most of us don't even know.

Just think of it: every single piece of technology developed from the 1990s onwards is largely based on one form of agile project management or another. Without agile, these products would have taken *a lot* more time to get to the market, and as such, our very lives would have been a little sadder without the existence of the internet, Wikipedia, and the infinite scrolls on Instagram every morning as we sip our cups of coffee.

Aside from the history and the advantages of agile, our first installment of this series also tackled a very important topic: how the basic Agile Principles (all twelve of them!) apply to real life and how they are translated into agile practices common across the entire spectrum of light-weight methodologies.

We know just how difficult the move from traditional waterfall approaches to agile project management can be. And, as such, we dedicated an entire chapter of our first book to showing you how to efficiently implement this framework into your business, regardless of what it may focus on.

We also tackled the topic of project management tools and provided readers with a brief analysis of some of the most popular tools one can use when embracing agile project management.

Finally, we dove deeper into two of the most complex subjects related to project management: risk management and scaling. The first topic is difficult to grasp in the context of agile project management because many people think agile involves no planning at all (which is highly untrue, as you will see in this book as well). The second topic is difficult because there are specific methods and best practices you should make sure to employ when aiming to scale agile, particularly in larger companies, where structures and documentation are already running through the very tissue of their organizational practices.

Enough about our previous book, though.

What does this book bring forward?

We mostly want to focus on the specific agile project management methodologies used by companies these days. While encompassing *everything* in the book is definitely not feasible, we aim to provide you with enough information to help you decide which of these approaches is best for your business.

We don't want to be repetitive, so we made sure to select the most precious bits of information for you. However, we do not claim the volume at hand to be a one-size-fits-all "recipe book." Au contraire: if you want this book to give you specific instructions on which method is the better choice, then you should probably put it aside.

We want to be completely honest with you (as we were in the first installment of this book). We cannot promise you the world and not deliver, and we genuinely believe that any kind of resource that will promise to give you the *ultimate* solution to anything is a complete and utter lie.

There's no such thing, regardless of whether you are talking about project management, apple pies, or efficient workouts.

Everyone is different. Every company is different. So, instead of chasing *ultimate solutions*, we encourage you to chase something more challenging but equally rewarding: *your* solution.

We want to empower you to make your own decisions when it comes to which agile methodology is best for you. As such, this book aims to give you the basic information on six of some of the most popular approaches out there: the definitions, the inner works, the principles, best practices, and all the other features that define them.

Our goal is for you to have the full picture of the agile project management world by the end of this volume. We want you to be able to analyze all the information we have provided both in this book and in our previous installment of the same series, and we want you to make the decision that works best for you, for your team, and, ultimately, for your organization.

As such, the first chapter here will be dedicated entirely to Scrum project management. As one of the two most popular methodologies, Scrum is frequently touted in mainstream business media as a receptacle of everything good in agile. While it is definitely full of advantages, Scrum should be perceived at its true value.

As such, we aim to provide you with a comprehensive overview of what Scrum is, how it relates to the general agile project management approach, how it connects to other agile methods, and how it functions. We will tackle Scrum teams, Scrum Ceremonies, Scrum Artifacts, and

Scrum rules, so that you can gain the bird's eye view on how Scrum project management is actually practiced.

Our second chapter will be all about Kanban and lean. As a method that originated long before all the other popular agile approaches these days, Kanban is an interesting approach in the world of agile—one that did not pertain to the software development industry to begin with but which has become an integral part of it over the past two decades.

We will analyze the main Principles behind Kanban and lean project management, as well as the best practices that make these two connected approaches so unique and so useful for businesses from such a wide range of industries.

Further on, we will also analyze Kanban in relation to Scrum, provide you with a list of benefits Kanban brings along, and show you how Kanban can be implemented with the help of a popular project management tool.

Next, we will move on to another popular project management approach, one that is more exclusive to the software development world than many others are these days: Extreme Programming (XP).

We will take a look at what XP entails, from the Planning Game to the Coding Standards and the importance of the Metaphor in the entire Extreme Programming paradigm. By the end of this chapter, we hope you will have a full view of what XP is and how to incorporate it in your agile approach (in its pure form or as a hybrid between other agile and nonagile frameworks).

Our fourth chapter starts a series of analyses on one of the less popular, but equally valuable agile methodologies out there: The Crystal

Method. You will be surprised to learn that despite the lack of popularity Crystal shows these days, it was one of the first ones developed in the 1990s (and it was also created within one of the most preeminent IT companies in the world!).

Beyond the debatable (lack of) popularity Crystal shows today, in 2019, this method has plenty to offer, its absolute flexibility being one of the most attractive features, even for larger organizations (like the one it originated in, actually).

Our fifth chapter in the book at hand is all about Feature-Driven Development (FDD), an agile method that might not sound very popular but which has been consistently incorporated with other agile approaches (including Scrum and Kanban). We will take a look at what FDD is and the elements that define it as a stand-alone methodology (elements that can be borrowed by other methodologies as well).

Last, but not least, we will approach one of the most apparently fearsome and misunderstood agile methods of the moment: The Dynamic System Development Method (DSDM). Frequently left aside because it feels too intricate and strict, DSDM can still provide you with plenty of value, particularly if you work in a company that is focused on structures and documentation and where a more relaxed agile approach would not function very well.

It would be remiss of us to think this book encompasses everything all these agile methodologies are. There is a very good reason there are so many books and so many materials and courses on these topics: they are wide and complex, much wider and much more complex than what many of the articles on Google would want you to think.

Truth be told, none of the methodologies we will approach in this book is actually *easy*, Agile itself is not easy, despite its apparent lightheartedness. Agile takes obsessive discipline, it takes a good strategist behind it, and it takes a lot of determination to function.

More than anything, agile takes a complete change in mindset. In agile projects, people do their jobs because it's the right thing, because this is what they are paid for, and because they want to be part of a successful story at the end of the project. They don't do it all because there's a "monster" called project manager that whispers "deadline" in their ear all the time.

In agile, things are done because they need to be done for the sake of evolution. People working in agile frameworks are more likely to have personal success stories in their résumé precisely because they get used to a mentality of hard work, dedication, and complete honesty. They know the path to growth is never easy and that it takes multiple increments to actually make it through, and they know this rule applies not only to whatever their job is but to their personal lives as well.

Agile has changed the way we think about success. It may not be fully apparent now, but if you look at all the inspirational business stories, at all the athletes and all the artists that won the game, at all the lovely stories of beauty and success the world has offered us, *agile* was always there.

They may not have called it that. In fact, it is quite likely that they didn't even think of a project management framework when they proceeded on their journey. But all these people you see on TV know that success is based on:

- honesty with oneself
- dedication
- hard work
- iterative and incremental implementation of growth
- constant feedback

At the end of the day, this is what agile is all about, regardless of whether you look at the oddities of Scrum and its Ceremonies or the apparent strictness of DSDM. And that is the main idea behind this book as well: there is no right or wrong agile approach. But some may just not be a good fit for you, and that's all right.

We hope the information lying ahead of you will not give you a headache. Instead, we hope it will open your mind (and, ultimately, your heart) to a world of opportunities in terms of what agile is (and what agile isn't).

Good luck and happy learning ahead!

Chapter 1: Agile Scrum Methodology

Agile project management is a complex topic, and one we have discussed quite in-depth over the course of our first book.

In (very) short, agile project management is all about keeping things flexible: from the way you plan ahead to the way you adapt to all the changes along the way. In turn, this helps deliver better products, in a shorter amount of time, and within a more acceptable budget.

As was shown in the first installment of this book series, when the official foundation of agile project management was laid back at the beginning of the 2000s, there was a real need for it. Software companies were experiencing massive lag in their deliveries of the final product precisely because the management system was not properly adapted to an industry where change is constant.

Agile project management has evolved a lot, and it has managed to incorporate a variety of directions meant to suit different types of companies, industries, and specific situations. Very often, organizations choose to work on hybrid frameworks (including hybrids that combine traditional/waterfall methods with agile methods).

Of all the agile frameworks out there, Scrum is, by far, one of the single most popular ones. This chapter will tackle the basics of the Scrum methodology, from what it is to how it connects into the 12 Principles of Agile, how the team is structured, what artifacts are, and how to actually practice it.

What is Scrum Project Management?

When they hear "Scrum," most people think of the (in)famous Daily Standup meeting. From the outside, it looks like a bunch of people sitting in a circle and taking turns in standing up and talking for two minutes.

It is easy to understand why this might seem silly, forced, or downright annoying.

The absolute truth is that Scrum project management is about a lot more than just the Daily Standup.

Indeed, this is a central part of Scrum project management, one that is not silly at all, actually.

Beyond that, Scrum project management has a lot of components and each of them brings its own contribution to the benefits of agile Scrum.

There are many ways to define Scrum project management (including through the perspective of the popularity with which it has been used). One of the best ways to do it is by calling it an "agile framework" that assigns specific roles to each team member, focuses on regular meetings at different levels, and focuses on a very well-defined mindset.

Scrum has gained a lot of fame in a relatively short amount of time. From the early 90s when it was first developed to today, Scrum has managed to become not only one of the most popular agile methods but very often *the main* method people associate with the entire concept of "agile project management."

The popularity of Scrum is largely due to its ease of use. Everything about Scrum is meant to put people in a certain productivity mindset, rather than impose on them specific tactics. Although from the outside it might seem the other way around, the techniques specific to Scrum are not meant to seclude the mindset of those who participate in a project, quite the other way around actually

Studies show that 94% of companies used agile Scrum project management as part of their approach in 2017. Only 16% of them used Scrum exclusively, while the remaining 78% used Scrum in combination with other methods.

Scrum project management is, without a doubt, one of the best ways to ease your way into agile project management, precisely because Scrum is very close to the ideal agile projection of how a product should be planned, developed, tested, and delivered.

Beyond any other definition, though, what Scrum is truly defined by are its benefits. As mentioned before, all agile methods were born out of a dire need: that of delivering faster, better results.

In (very) short, these are some of the benefits of Scrum project management:

- Generating revenue. Because Scrum involves an incremental release of the product, it is far more likely that money will come back into the business sooner, rather than later.
- Providing better quality. The incremental development of the product ensures that it will be perfected with each iteration. As such, the end product will be of higher quality as well.

- Offering better transparency for everyone. The entire process is transparent for every stakeholder and team member, from beginning to end.
- Better risk management. The incremental nature of Scrum also helps Scrum Masters/project managers manage the associated risks with a lot more efficiency, as they will adapt the risk management plan along the way, according to the challenges they encounter.
- Better flexibility. Since this is agile (which means it is a project management approach that is both iterative and incremental), you will have better flexibility to adapt to changes.
- Better control of the costs. Again, this is due to the incremental nature of Scrum and how it allows you to manage the entire project more accurately.
- Better customer satisfaction. This ties back into the fact that Scrum allows you to release *chunks* of product and attract feedback from your customer. As a result, your final product will be better as well.
- Faster product release. When you can adapt the product along the way, you can actually release it sooner too.
- Better work environment. Scrum promotes a culture of cooperation, collaboration, communication, and a good work/life balance. All of these things lead to a better work environment, where people are allowed and encouraged to grow.

What is Scrum in Relation to Agile Project

Management?

Scrum project management is so tightly connected to the world of agile project management that the two have become almost synonymous in the collective mind.

Agile and Scrum are so tightly connected that when someone says "agile," they almost instantaneously think of practices that are very specific to Scrum, like the Daily Standup we mentioned before, the Scrum Master, or the Scrum board, for example.

Indeed, if you look at the benefits of general agile project management and Scrum project management, you will see how close they are in nature. In most cases, the latter is just a practical, "regulated" mirror of the philosophy the first tries to bring forward.

Even with all the closeness between the two concepts, it is still very important to make sure you understand that they are quite different.

If you want to picture it, imagine a tree that makes apples of different colors. The tree is the agile project management approach, while the colored apples are the different methodologies (which we will approach throughout this book). Scrum is one of those apples. While it grows from the same tree and follows the same general rules on how apples grow, it has chosen to have its own unique color.

Agile project management is an approach that focuses on iterative development, while Scrum is a branch that has created its own rules of how to play the game by the general Agile Principles as they were laid down in 2001.

In the end, you cannot actually compare agile project management and Scrum. One is an entire category of approaches, while the other one is a specific approach. Indeed, it is an approach that gets very, very close to the ideological and methodological basis of agile project management.

However, when you discuss agile project management and Scrum in relation to each other, you should do it from the perspective of evolution, growth, and branching, rather than from the perspective of comparison per se.

What is the Scrum Methodology Compared to Other Agile Approaches?

The world of agile is quite large and fascinating, and, unlike the world of waterfall, it is a world full of nuances.

The vast majority of companies don't take *one* methodology to heart. Rather than that, they create their own mixes, according to their specific needs and their specific company policies.

Scrum is, as it was mentioned before, one of the most popular agile methodologies, so much so that it is almost synonymous with "agile" in general. However, that doesn't make the other methodologies less agile; it just makes them different.

At their foundation, all agile methodologies spring from the 12 Agile Principles, but each of them shows its own particularities.

For instance, when you discuss Scrum and Kanban, you might feel that you are discussing the same methodology. In reality, however, they are quite different in how they both approach and tackle the Agile Manifesto's dictums.

The differences between Scrum and other agile methodologies will become clearer as you move through this book. However, let us briefly dive into a short comparison that will give you a better idea on how Scrum relates to the other popular agile methodologies.

Scrum vs. Kanban

As mentioned above, the two are quite similar in the sense that:

- Both abide to the general agile Principles.
- In both cases, the work in progress is limited (but Kanban emphasizes more on this).
- In both cases, the work is broken down in small increments.
- In both cases, work is scheduled using the "pull" method, rather than the "push" one.
- In both cases, the teams are organized.
- Both methods emphasize transparency.

As for the differences between the two, they include the following:

- Scrum uses time-boxed iterations, while Kanban doesn't.
- Scrum is all about fast-changed processes, while Kanban isn't.
- Scrum uses burn-down charts for each iteration, while Kanban doesn't.

- Scrum limits the work in progress through its sprint plan, while Kanban limits it through its workflow (each team member is allowed to handle only one task at a time).
- Kanban is generally less structured.
- Scrum assigns prescribed roles (Product Owner, Scrum Master, etc.).
- Scrum is all about getting work done faster, while Kanban is all about improving the process.

Scrum	Kanban
Timeboxed iterations prescribed	Timeboxed iterations optional. Can have separate cadences for planning, release, and process improvement. Can be event-driven instead of timeboxed.
Uses Velocity as default metric for planning and process improvement.	Uses Lead time as default metric for planning and process improvement.
Cross-functional teams prescribed.	Cross-functional teams optional. Specialist teams allowed.
Items must be broken down so they can be completed within 1 sprint.	No particular item size is prescribed.
Burndown chart prescribed	No particular type of diagram is prescribed.
WIP limited indirectly (per sprint)	WIP limited directly (per workflow state)
Estimation prescribed	Estimation optional
Cannot add items to ongoing iteration.	Can add new items whenever capacity is available
A sprint backlog is owned by one specific team	A kanban board may be shared by multiple teams or individuals
Prescribes 3 roles (PO/SM/Team)	Doesn't prescribe any roles
A Scrum board is reset between each sprint	A kanban board is persistent
Prescribes a prioritized product backlog	Prioritization is optional.

Scrum vs. Extreme Programming (XP)

Although XP is probably less popular than Scrum (and Kanban, for that matter), it is worth looking into how it relates to it.

Some of the main similarities between the two methodologies include the following:

- They are both agile.
- They both focus on delivering quality products in a short time span.
- Both start with a planning meeting.
- Both use sprints to organize the work.

As for the differences, they include the following:

- Scrum includes Daily Meetings, while XP doesn't.
- The length of each sprint is usually shorter in XP.
- The focus lies on delivering working software, rather than doing it by a specific product release date.
- Scrum doesn't allow changes to be made during a sprint, while XP does.
- XP focuses more on engineering principles (such as automated testing, test-driven development, pair programming, and so on).
- In Scrum, the meetings are coordinated by the Scrum Master, while in XP the team members take turns in coordinating meetings.

Scrum	XP
• Changes in sprint are not allowed • Once tasks for a certain sprint are set, the team determines the sequence in which they will develop the backlog items • The Scrum Master is responsible for what is done in the sprint, including the code that is written • The validation of the software is completed at the end of each sprint, at Sprint Review	• As long as the team hasn't started working on a particular feature, a new feature, of equivalent size can be swapped into the interation in exchange for an un-started feature • Tasks are taken in a strict priority order • Developers can modify or refactor parts of code as the need arises • The software needs to be validated at all time, to the extent that tests are written prior to the actual software

Scrum vs. Crystal

Scrum and Crystal are also worth comparing. In short, the main similarities between the two include the following:

- Neither puts a lot of emphasis on documentation.
- Both abide by the general agile Principles.

There are some more than notable differences as well:

- In Scrum, the client is represented by the Product Owner, while in Crystal, the customer is involved in the process.
- Crystal does not connect planning and development to specific requirements like Scrum does.
- Crystal focuses even more on face-to-face meetings.
- Crystal focuses more heavily on the *team* members and adapts the processes according to them.

Scrum vs. Feature-Driven Development (FDD)

Feature-Driven Development (FDD) is still one of the most extremely valuable agile methodologies out there. Although different in many ways, it does show some similarities with Scrum, such as:

- They are both agile methodologies.
- Both of them follow the same basic steps.
- They can act as a complement to each other.

The differences between FDD and Scrum include the following:

- In FDD, there is no Scrum Master, but chief programmers who act as leaders and mentors.
- FDD is "ultra-light" as compared to Scrum, as it is far less prescriptive and more adaptive.
- Scrum does not recommend any specific engineering practice.
- FDD shows a longer feedback loop.
- Scrum focuses on self-organized teams more than FDD.

Scrum vs. Dynamic System Development Method (DSDM)

Same as with the other methodologies we have touched upon in this comparative section, Scrum and DSDM show both similarities and differences.

Some of the most important similarities between Scrum and DSDM include the following:

- Both of them abide by the general agile Principles.
- Both of them can work when they are combined.

As for the differences, the most notable ones include the following:

- DSDM tends to be even more prescriptive in terms of team roles.
- In the case of DSDM, all the basic information has to be set from the beginning: features, quality, time, and cost.
- DSDM tends to be easier to adapt to a corporate environment.

This is, of course, a very brief overview of how Scrum and other agile project management methodologies are similar and different at the same time. As we go more in depth on each method, the similarities and the differences will become more obvious.

The Scrum Team

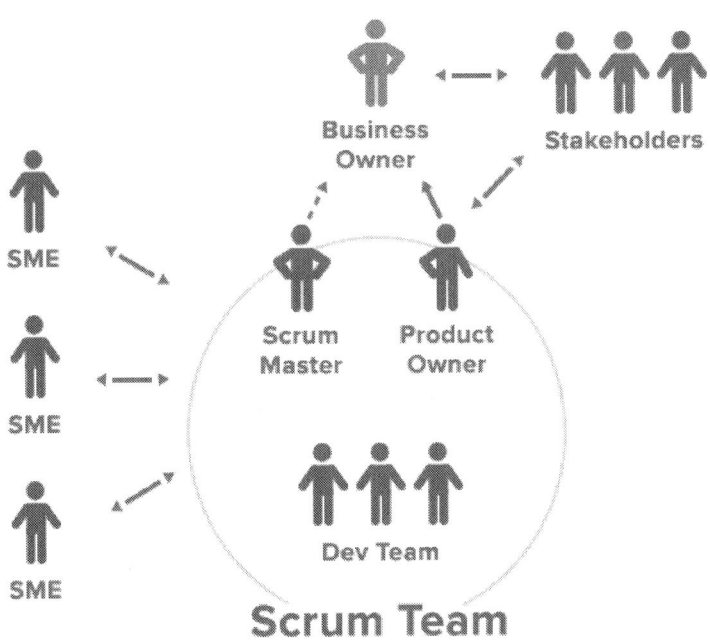

Scrum Team

The Scrum team lies at the very core of the entire methodology. It is where Scrum starts and where the true magic happens too. It is also one of the Scrum features everyone from the outside can easily spot (let's face it, adding Scrum Master to your resume sounds way cooler than other titles, doesn't it?).

Beyond the buzz and the "glam" of what the Scrum team feels and looks like from the outside, it is more than worth noting that in a Scrum team, everyone has their own part to play. Everyone's role is quite clearly defined, and everyone comes together to fight for the common goal:

delivering the project on time, within the limits of the budget, and at the highest standard of quality possible as well.

In general, a Scrum team will be split as follows:

The Scrum Master

The Scrum Master's main role is that of planning and making sure the plan is followed through. Most of the times, this means he/she will have to:

- clear obstacles
- create an environment of efficiency
- address team dynamics
- create and maintain the communication and relationship between the Product Owner and the team
- protect the team from external interruptions

The Product Owner

At the confluence between a Scrum Master and a more traditional project manager, the Product Owner in a Scrum team has a pretty well-defined role as well. They are meant to:

- act as a liaison between the Team and the Customer
- act as a liaison between the Team and other Stakeholders
- help the Team estimate the size of the stories
- help the Team split larger stories into smaller chunks

The Team Members (Development Team)

This part is pretty straightforward: The Development Team is a group of people handling the actual development of the product. Some of these people might act as developers and others might act as Quality Assurance engineers, for example. But together, they form the "Team" in a Scrum environment, and together with their Scrum Master, they will coordinate themselves to create products that suit the user requirements and the information collected by the Product Owner from the customer.

Aside from the roles held in a Scrum team, it is also more than worth mentioning that there are a few guidelines all Scrum teams are meant to follow (and all of these guidelines are inspired by the agile Principles, of course).

Some of the most important rules in a Scrum team include the following:

- Everyone follows the same rules and works for a common goal.
- The Team (in its entirety) must be accountable for the product delivery.
- The Team must be empowered to work at its maximum efficiency.
- The Team must be autonomous and self-organized (as much as possible at least).
- The Team is usually quite small and there are no subteams under it.
- There must be a balance of skills within the team.

Depending on how "purely agile" and "purely Scrum" you want to be in your approach, you might also want to make sure that your team is located in the same place, so that meetings and collaboration can take place face to face every time.

Scrum Events (Ceremonies)

Aside from the predefined roles, Scrum also allocates very specific events (also called "Ceremonies") for very specific purposes.

There are four main Scrum Ceremonies, each with a very clear goal, as follows:

Sprint Planning

The Sprint Planning Ceremony is all about, well, *planning*. Since the entire project will be split into small iterations (usually lasting for a couple of weeks or a little more), every such iteration will start with a new planning session.

As such, a project will have as many Sprint Planning sessions as there are sprints. Furthermore, every Sprint Planning will last for about one or two hours, depending on how many issues there are to discuss.

It is of the utmost importance to include the entire team in every Sprint Planning, so that everyone can participate in the planning itself. This way, everyone will bring their own input to the table, and you will be able to create a more accurate plan for the Sprint ahead of you.

The Scrum Master, the Product Owner, and the Development Team all have to participate in the Sprint Planning if things are to be handled smoothly from thereon. DO encourage people to communicate if they are skeptical or doubtful about anything!

Daily Scrum

This is the most well-known Scrum Ceremony *ever*.

Almost everyone has heard about it.

But the truth is that most people completely misunderstand what the Daily Scrum is all about.

In short, the Daily Scrum is a short, 15-20-minute meeting held every day. The Scrum Master and the Development team are the main participants here. They gather in a circle and each of the members takes their turn standing up and talking for 2 minutes about:

- what they did the day before
- what they plan on doing today
- what bottlenecks they might have

Despite what many would believe, the Daily Scrum has nothing to do with micromanagement and constantly following your team to ensure they do their job. In fact, the Daily Scrum is meant to help team members be more productive by removing the bottlenecks from their path to achieving what they aim for every day.

Sprint Review

In this Ceremony, the Team demonstrates the work they have done throughout the duration of the sprint.

The Sprint Review does not have a time cap; it can last for as long as it is needed for the team to demonstrate the work, they have done throughout the sprint that ended.

Furthermore, keep in mind that it is quite important to maintain the Sprint Review as positive as possible. Even if mistakes were made and even if the sprint results are not what you aimed for, it is still essential to congratulate the team for what they did right. This will help you maintain team morale, and it will help you step into the new sprint.

Sprint Retrospective

If every sprint starts with a Sprint Planning session, it must end with a Sprint Retrospective as well.

This is a time to bring the entire Team together and analyze what went right and what didn't go as right as it should have. It is a time to discuss, but it is important for these discussions to lead to actual action too.

Like all agile methodologies, Scrum is all about adapting to change, and the Sprint Retrospective is meant to help the Team incorporate the changes they need to make for the future sprint.

Scrum Artifacts

Aside from the predefined roles and the Ceremonies, Scrum also makes use of so-called "artifacts."

What these artifacts are, in fact, are tools used for better planning and development of a product. Although quite specific to Scrum itself, the most common artifacts in this agile project management methodology have been borrowed by other methodologies as well, precisely because they are so useful.

The most popular Scrum artifacts include the following:

Product Vision

This artifact describes the product in a nutshell. It should be something short and easy to remember—something that gives everyone a good direction and helps remind them of the higher goal behind their daily work.

Sprint Goal

This tool is used to describe the specific goal of each sprint. For instance, if you are developing a social media management platform, your first sprint's goal might be to create the basic platform upon which all the features will be built (i.e., a platform that connects to various social media channels).

Product and Sprint Backlog

These two artifacts are very frequently mistaken and considered to be one and the same, but they are quite different in essence.

In short, a product backlog is the entire collection of tasks that have to be handled throughout the development process (until the delivery of the final product). A sprint backlog will subtract those tasks that pertain to that specific sprint only, though.

Burn-Down Chart

This artifact is a chart the Scrum Master and the Team will use to observe the progress of the team's efficiency, as well as how realistic the initial plan was, over the course of the development process. The burn-down chart will be used to make adjustments on the go, as needed.

Increment

To define it simply, an increment is a Scrum artifact used to describe the totality of the Product Backlog items completed during a sprint (as well as all those that were completed in all the Sprints before it).

Aside from these artifacts, it is also very important to remember the fact that you absolutely *have to* set the definition of "done" with your team members. Believe it or not, people have different versions of what "done" actually is, so it is essential to clear things up with your team before you even start working.

Scrum Rules

The basic rules of Scrum are, in fact, the basic rules of agile. As long as you abide by the 12 Principles (as they were extensively described and explained in the first increment of our Agile Project Management book), and as long as you make use of the Scrum artifacts, Scrum Ceremonies, and Scrum roles, you can definitely call yourselves a Scrum team.

Aside from all that was mentioned so far, it is also worth noting some of the most important rules in Scrum project management:

- A sprint cannot last for more than four weeks (ideally, it will last for less than that).
- You shouldn't take breaks between the different sprints.
- You should make sure all sprints last for the same amount of time.
- The main underlying goal of every single sprint should be "workable" and potentially "usable" software that can be delivered to the customer as such.
- You shouldn't skip any of the Scrum Ceremonies (the Planning included).
- Most meetings should be time boxed, except for the Sprint Review.

- Your Daily Meeting should take place every day, at the same time of the day, and preferably in the same space.
- Sprint Reviews should include feedback from the stakeholders at all times, so that you know what conclusions to draw during the Sprint Retrospective.
- There should be no break between the Sprint Review and the Sprint Retrospective. The feedback received during the Review should be fresh for the Retrospective, so that the entire team can discuss it and jot down the actionable steps to implement in future Sprints.

Practicing Scrum

In theory, Scrum sounds like actual fun because who doesn't like planning with their entire team, and who doesn't like not having to constantly draw attention to each and every member of the team to do their work?

In practice, Scrum might prove more difficult to implement, especially in a business that is already used to traditional project management methods. It all depends on how smoothly you want to introduce Scrum to your team and your higher management alike.

Some of the best practices you might want to incorporate in your approach include the following:

- Stakeholders should be an integral part of the entire development process. Invite them to some of your meetings, particularly Sprint Planning and preproject meetings.
- If you already have a team that is well formed, don't break it. It is far easier to introduce Scrum to them because they already know each other, they already speak each other's "language," and they already know how to properly communicate with each other.
- Team building is important. It doesn't matter if you choose Scrum, another agile methodology, or simply choose to remain traditional. Team building can really make a difference.
- Include the team in the Sprint Planning, especially when it comes to making estimates. We have already tackled the subject of how to play estimation "games" with your team in our first installment of this book series, so we definitely advise you to check that out. These games are not just fun; they also help you get actually better estimates of how long some tasks will take until completion.
- The Product Backlog and the Sprint Backlog should be delimited for better organization.
- Try to prioritize the items in the Product Backlog according to their importance and how dependent other items are on them.
- DO use a Scrum board. Typical columns in a Scrum board will include: Stories, Not Started, In Progress, Done, Blockers, In Testing, In Review (by the Product Owner, for example). These columns should be considered as separate from the Product Backlog and the Sprint Backlog.

- Calculate the velocity of your team with regularity. It will help you spot any efficiency problems while they are still young and easily repairable.
- Your developers and your testers should work together, and yes, this means they should be in the same room as well.
- All the bugs you find during a current sprint should be fixed in the following sprint.

Obviously, this is the very short version of what the best practices of Scrum look like. As mentioned at the beginning of this section, though, it is far more important to adhere to the general Principles than any prescriptivist methods.

Scrum can be amazingly useful, and it can definitely be a very good "gateway" into the world of agile in general. Because its practices and its particulars are much more popularly known among many people, your team and your higher management might be more open towards adopting Scrum (as opposed to other, more extreme, and far more lightweight methodologies).

Scrum tends to be a milder version of agile, and for this reason, it also tends to be combined with practices from other agile methodologies (as it was also mentioned when we compared Scrum and the other approaches we will tackle throughout this book).

We invite you to learn more about the other agile methodologies as well. Next chapter, we will discuss Kanban (the other, probably equally popular agile approach). Following that, we will dive a little deeper into the world of agile and take a closer look at XP (Extreme Programming), Crystal, Feature-Driven Development (FDD), and the Dynamic System

Development Method (DSDM), all of which are incrementally more agile, and all of which pose particulars that might be more suitable for your company.

In the end, you and your team will decide which of the methodologies works best for you or if you need to combine multiples of them to create the right "program" for your team's best efficiency.

Chapter 2: Lean and Kanban Software Development

Kanban is, perhaps, the oldest project management method in the realm of agile. While it might have been officially adopted as part of the agile family when the Principles and the Manifesto were laid down, Kanban's history goes far beyond that.

The beginnings of Kanban date as far back as the 1940s, when Toyota designed the first Kanban system. Albeit far more rudimentary than what is nowadays known as the "Kanban agile project management methodology," the system created by Toyota back then survives to this date, and it has been borrowed by and adapted to a multitude of industries.

What started off as an approach that meant to minimize waste and improve efficiency is, these days, an entire methodology used in software programming, marketing, hospital management, and a variety of industries that are apparently as disconnected from each other as they can possibly get.

When you get beyond the differences between all these industries, though, you realize that all of them face the same kind of issues that pose the same kind of questions:

- How to reduce waste?
- How to maximize the efficiency of a team?
- How to make the work environment a healthier and more productive one?

A bit less prescriptive in nature than Scrum, Kanban manages to offer an answer to these questions (and more). As you will see throughout this chapter, Kanban and lean project management are very tightly connected, and they both fit under the agile project management umbrella quite well.

Without further ado, let's take a closer look at what Kanban is and how to use it in your company.

Main Principles of Lean Methodology

The lean methodology (or "lean thinking" as it is sometimes referred to) is not necessarily a project management approach, framework, or methodology in the fullest sense of the word.

Rather, this is a business philosophy that is adjacent to many of the agile project management methods out there.

The lean methodology relies on the Japanese concept of "Kaizen" (meaning "improvement"). So, right from the very beginning, you can see how tightly connected lean and agile are in essence. In its turn, Kaizen relies on five main principles:

Value

In Kaizen, providing value to the customer is the ultimate goal. However, in order to do that, you will first have to understand what "value" means for each of your customers. What makes them pay for your product, regardless of whether that is a car, a software program, or a piece of clothing? What triggers someone to buy the product and be happy with it?

Once you understand the kind of value you can offer your customer, you also understand how to minimize waste and how to create a product that is efficient both in terms of quality and in terms of price.

Value Stream

The Value Stream is the road map of the entire product, from idea to the end user. If you want to eliminate waste and provide your customer with genuine value, you have to understand every single step the product will take, from its raw materials to how the customer will use it.

Flow

In lean, your manufacturing/development flow is of the utmost importance because you have to constantly make sure there are no pauses in production. Every pause is a waste of time, and since eliminating waste is the beating heart of every lean approach, it is unpardonable not to ensure a smooth flow of the entire production process.

Pull

Eliminating waste also means that you will have to create a number of products that is *just enough*. Too much product will lead to waste, but a small amount of product will not satisfy the customer. As such, lean project management makes use of the *pull method* to ensure the production process is just in time and that the product (and number of products, in the case of the physical ones) is perfectly coordinated with the demand (of the market or of a specific customer).

Perfection

Same as in agile project management, Kaizen promotes the idea of *continuous improvement*. Its end goal is nothing less than perfection: a product that works perfectly, comes to the market at the perfect time, and fits the needs of the customer in a perfect way.

While absolute perfection might be hard to reach (if not impossible, in most cases), striving towards it will help you and your team *ask for more* from yourselves and continuously work towards an ideal goal.

Aside from the Kaizen principles that support lean project management, there are seven more principles that come as a complement to everything and circle back to the idea of providing true value to the customer. What is commonly known as the "Seven Lean Principles" includes the following:

- eliminate all waste (as much as possible)
- support and encourage learning and knowledge
- decide as late as possible
- deliver as soon as possible
- emphasize the power of the *humans* in your team
- believe in integrity and quality
- constantly optimize and observe the *whole*

These Seven Principles will be discussed throughout the next seven subsections of this chapter, so we will not dwell on them too much. It is important to know, however, that they connect very closely to both the Kaizen Principles and to the 12 Principles of Agile project management. Looking at the lean principles through these two points of view will allow you to understand where Kanban falls on the agile spectrum, and, ultimately, whether or not it is a good choice for your team and your organization.

Eliminating Waste

7 Wastes of Lean

Inventory   Waiting   Defects   Overproduction

Motion   Transportation   Over-processing

In lean project management, waste is the ultimate evil. You will never find an agile methodology to focus so much on this specific aspect of production, and this is mostly related to the fact that lean and Kanban both draw their roots from manufacturing environments, where waste is very palpable, physical, and, well, *costly*.

You might not be able to sense waste in software programming, not in the sense that you can hold it in your hand or that you can look at a pile of it as it collects in the corner of your office. You can, however, *feel* its effects, and this is where lean proves its efficiency in everything connected to software project management.

In the original form of lean and Kanban, the precursors of this approach identified eight main types of waste one should avoid at all times:

1. Transport
2. Inventory
3. Motion

4. Waiting

5. Overproduction

6. Overprocessing

7. Defects

8. Skills (also known as the 8th type of waste because it was not originally included in the core principles of lean)

These types of waste are quite self-explanatory when you look at them from the point of view of a manufacturing company. However, let's put them in the context of a software development company, so that you understand how they can still apply even when the final product is not a physical one:

1. Transport may refer to an unnecessary movement of information (e.g., too many reviewers involved in the process, too many backs and forth emails passed between the different stakeholders, and so on).

2. Inventory may refer to accumulating too much information to be entered in the system at a later date or accumulating too much information that is not related to the other members of the team in due time.

3. Motion may refer to switching too many tools, moving too much between different workstations, and, in general, *moving* too much.

4. Waiting may refer to any kind of bottleneck that prevents one (or more) of your team members to move further with the development process.

5. Overproduction may refer to anything that ranges from hitting the "Reply All" button (when not everyone is interested) to simply delivering *much* faster than needed.
6. Overprocessing may refer to adding any kind of extra feature that was not requested initially.
7. Defects may refer to any type of bugs or issues the product might have.
8. Skills may refer to a lack of talent in your company (e.g., you are struggling to find senior developers).

As you can see, the same waste principles can be applied to manufacturing just as they can be applied to software programming. Likewise, you can transpose the same principles to any other industry: governmental institutions, educational institutions, health organizations, and so on.

Waste is a true demon, especially in a world that seems to be infatuated with wasting time, food, and resources on a daily basis. Fighting off this demon might not be easy, but as the decades of experience at Toyota prove, it is more than worth the effort.

Amplifying Learning

The basic concept behind this principle is quite simple: the more you and your team learn and build knowledge, the more efficient the process will be and the better the final product will be on all verticals (cost, delivery time, quality).

Just because the principle itself is simple to understand, it doesn't mean it is equally simple to apply in real life. In theory, acquiring knowledge always sounds easier than what it actually is in practice, and this is why you have to pay attention to *how* exactly this principle is brought into materialization.

Some of the main ways to amplify learning in an agile Kanban team include the following:

1. Constant feedback. Since this is one of the main tenets of agile project management in general, constant feedback should be seen as an absolute must. Both you and your team must get used to always giving and receiving feedback from each other and from your customer. Most importantly though, you all need to make sure you actually *apply* that feedback as well (there is no point in giving and receiving feedback if you don't do anything about it).

2. Asking yourself "how" and "why." When you want to get down to the root of a problem, these two questions will lead the path. Imagine them as your guiding light through the harshest moments of your project. Do keep in mind that it is important to be honest about your answers. Also, do keep in mind that you should go in-depth with your answers as well; if needed, ask the same question multiple times to get to the root cause of the problem.

3. Pair Programming. Although this is not a practice inherent to Kanban, it is a useful exercise for your team members. Basically, what Pair Programming means is pairing your developers in groups of two. One will be the driver (the one who writes the

code), and the other one will be the observer (the one who guides the driver).

4. Code reviews. Also known as "peer code reviews," this practice will help your team members give feedback to each other and learn from each other's experience.

5. Documentation and wiki. Although agile project management doesn't emphasize the importance of documentation, it doesn't mean there should be *none*. The focus should lie on working and delivering the product, of course. However, documentation and wikis (knowledge bases) can be passed on from one team member to another (including new recruits who might not be completely familiar with the processes and practices you have employed so far).

6. Training sessions. Training, courses, and seminars will definitely help everyone grow. It is quite important to pick your subjects and your training carefully. You don't want to bore your team with information they have heard a thousand times before. Instead, you want these sessions to prove genuinely helpful for everyone.

7. Track your progress. Charts and project management tools can make a world of difference when it comes to tracking the progress of your team, both in terms of efficiency and in terms of quality.

These are some of the most common practices you can employ to accelerate the learning of your team. As mentioned in the beginning, the more they know, the more they will be able to deliver in terms of quality, efficiency, and speed.

Deciding as Late as Possible

This is one of the most poorly understood principles behind lean and Kanban. A lot of people believe this principle refers to postponing all decisions until it is too late (and thus, displaying an irresponsible behavior towards decision making).

In fact, this principle is about the exact opposite concept: keeping your options open. Making decisions too early in the process might affect the end result if you don't have all the data at hand. However, postponing the final decision until you have plenty of data to make an informed decision will help you get closer to what really needs to be done.

In other words, there is no point in planning for months in advance. As per the agile Principles, you should always be ready to adapt to change, and a rigorous, strict plan will not help with that.

This doesn't mean you shouldn't plan *at all*. However, you should do it in smaller increments, so that you can easily adapt along the way as per the information you receive throughout the development process.

Delivering as Fast as Possible

Again, this is a frequently misunderstood concept, not just in lean and Kanban, but in agile project management in general. Delivering ahead of the schedule is one thing, but making a bad job out of your delivery is a completely different matter.

Delivering *fast* in agile (Kanban included) is not about sacrificing the quality of the product. Instead, you should focus on delivering *as fast as possible* while still maintaining the quality of the product at a high standard.

Instead of emphasizing how to be *faster*, emphasize on removing obstacles that might make you slower, such as:

- planning too far out in advance (especially for requirements that haven't been made yet)
- bottlenecks that have not responded to change
- going overboard with your engineering solutions

There is no recklessness in lean. Au contraire, this entire methodology focuses on being fully responsible and fully accountable for your own actions. Instead of rushing a final product to delivery, start with a basic product, deliver it, get your feedback, and then develop incrementally according to the aforementioned feedback.

Empowering the Team

Lean and Kanban are human-centric, meaning that they place the team members at the core of the entire process. This translates into multiple practices and approaches to work, product delivery, and, in the end, into the way your team feels about their work.

Empowering the team means putting them at the center of your methodology. Processes can be changed, products can be adapted to feedback, and budgets can be restricted. But people are rarely actually replaceable, and this is not related solely to the dire situation of the software programmers' job market but to the very dynamics of your team.

Once your team has been shaped, it means you have already invested time and effort into bringing everyone together and working with them. Losing any of the team members is a waste of time, of human resources, and of credibility (in the eyes of the other team members).

If the previously mentioned lean principles are misunderstood, this one tends to be downright forgotten. It is quite easy to understand why people might forget about the power of the humans in a team, especially in highly competitive, fast-paced industries (like the software development ones).

When you focus on delivering fast and on doing a good job, it is very easy to forget about the people who make this happen. Sometimes, it's easy to say *please* and *thank you*. Other times, it's easy to forget that there is life after work (both for you, as the project manager, and for your team members too). And, ultimately, it's easy to forget that adaptability to change is the primordial quality of all agile approaches, so everything can be turned around for a better perspective.

The only exception to this is the *humans* powering your development. When they are burned out, frustrated, and overworked, they will either disengage or call it quits. If you work in software programming, this is especially likely, since a good software developer will most likely find another job in a matter of days or weeks. For you, as the representative of your organization, it might take months before you find the right fit for your team.

Building Integrity In

Like it or not, quality isn't born out of nowhere, and yet, it is one of the core Principles of agile project management in general.

You can do everything right in terms of planning and you can hire the best people, but if you don't build a culture of integrity, self-organization, and responsibility, you risk the very quality of the final product.

Some people tend to take agile methodologies (Kanban included) as too *relaxed*. It's easy to understand why someone looking from the outside might see things as such. Agile teams are frequently pampered with relaxation rooms and PlayStation sessions, table tennis championships, and all the good coffee in the world. Looking at this from the outside, it might seem like the people behind the glass aren't even working.

Beyond all the glam and buzz of working in an agile team (particularly an agile software development team), all of these freedoms and benefits are allowed because there is full trust in each team member's ability to manage their time, their tasks, and their deliveries.

Integrity is a key principle in Kanban (and all agile project management methods, really). And you cannot have integrity if you don't have self-discipline and the genuine desire to *do* better.

Some of the exercises you might want to incorporate in your project management method to promote a spirit of integrity and commitment include the following:

- Pair Programming. As we have mentioned before, Pair Programming can help team members learn from each other and grow as professionals. It can also help them become more self-resilient, more capable of taking responsibility, and more devoted to true honesty.
- Test-Driven Development. Code criteria will help your developers write code that matches the business requirements and objectives. As such, it will help them stay true to a set of basic rules from the very beginning.
- Constant feedback. Again, this has been mentioned before, but we will mention it here as well for a very simple reason: it actually does help people become more responsible and honest about their own work and their skills.
- Maintaining focus. Although this is not an exercise per se, maintaining the focus of your team members on what they actually have to do will help them be more honest about the ways in which they manage their own time. Reduce all context switching and distractions so that your developers and QA engineers can focus on their actual job.
- Automate dull tasks. There's no point in making your developers waste time with copy-pasting and data entry when you can automate these processes. Also, try to automate pretty much any process that might be prone to human error.

Integrity and devotion aren't easy to build. But once you have set the foundation, they will help your entire project run more smoothly, so that you can deliver faster, better results.

Seeing the Whole

Splitting large projects into smaller iterations and increments is a basic tenet of agile project management in general, and Kanban makes no exception, as has already been touched upon.

However, one of the major risks associated with this type of approach is related to the fact that you can very easily get lost in the myriad of smaller tasks and lose focus of what is actually important.

As a rule of thumb, this is what the Kanban board is for, to allow both you and your team members to take a good look at the big picture every now and again. Furthermore, having a very clear and concise project goal can also help you stay on track and make sure you are all aiming for the same end result.

In software programming, suboptimization becomes a real issue, and it is crucial that you try to avoid it as much as you can. What "suboptimization" refers to is focusing on one or two indicators of success instead of the "whole."

For instance, you might be more tempted to focus on releasing low quality code for the sake of speed. When programmers are more or less forced to *deliver* whatever happens, they might become sloppy, and as such, the quality of their work might drop.

Short term, this might look like a solution because it will allow you to stick to one of the verticals of your plan: timely delivery. However, doing this will inadvertently have a massive impact on the quality of the product. Sooner or later, the same programmers will have to run through the code one more time to fix all the issues, and this will take a lot more time than if they would have done it right to begin with.

According to the lean and Kanban Principles, you should focus on the entirety of the project. Time, quality, and costs are all part of the "whole," and none of them should be ignored for the sake of the others. In other words, it is always best if you take the time to understand the vicious cycles your team might be prone to, so that you can come up with a better approach, one that will not sacrifice any of the major elements of a successful project.

At the end of the day, optimizing the whole is all about the value stream and how multiple elements come together to deliver a product as close to perfection as possible. Once you have identified the value flow in your team, you will be able to make the right decisions about a multitude of factors that might influence the end result, including:

- how to organize your team
- how to work with colocated and remote teams
- how to cope with inefficient team members

The whole point of this Kanban principle is to ensure that you don't get lost in the details. At the end of the project, your customer won't judge you by the number of *fine* iterations you have delivered, but by the quality of the product, its timely delivery, and how much you stuck to the initial budget.

How is Kanban Different from Scrum?

Items	Scrum	Kanban
Roots	Arose from software development.	Arose from manufacturing domain.
Framework	Within agile framework, which has the core values and principles.	It is actually from Lean production, though now used highly in Agile development.
Roles	There are 3 primary roles in Scrum: 1)Product Owner 2) Development Team 3) Scrum Master	Kanban does not have any specific role. You to start with the existing roles in the organization.
Iterations	Each iteration in scrum is called a sprint. (Usually from 2 weeks to 4 weeks)	There is no concept of iterations in Kanban. It is based on flow based mechanism.
Ceremonies/ Events	Four ceremonies: 1) Sprint planning 2) Daily scrum 3)Sprint review 4)Sprint Retrospective	No defined events.

But team can have its planning meetings or retrospective meeting as needed. |
| Prescriptiveness | Prescriptive, but not a heavily prescriptive process | Less prescriptive, as does not say which meeting to conduct, which roles to do. |
| Work in Progress (WIP) | In Scrum it is pull and the WIP is per iteration.

WIP term is not mentioned in Scrum. But, once product backlog items are committed into the sprint, they should not be changed – hence implicit WIP. | In Kanban, it is pull and the WIP is per workflow state.

WIP in Kanban is explicit. The limit is clearly defined and written on top of the columns of the board. Kanban has focus on flow, WIP, batch size and queues. |

In many ways, Kanban and Scrum are the ultimate "beginner agile methodologies." They are structured and prescriptive enough to make sense even for someone coming from a highly traditional environment. Yet, they adhere to the agile Principles so well that you simply cannot place them elsewhere.

At their very core, Kanban and Scrum are very similar, so it makes all the sense in the world that they have even given birth to a whole new "offspring" in the world of agile: Scrumban.

The similarities between Scrum and Kanban have already been relayed in the first chapter of this book and so have the differences.

If we have to nail it all down to just a few concepts, keep in mind the following:

- Both Kanban and Scrum are agile methodologies.
- Kanban was born earlier than Scrum and it was initially applied to the manufacturing industry.
- Both Kanban and Scrum use boards for visualization.
- Scrum teams are more structured than Kanban teams.
- Kanban puts more emphasis on waste reduction at all its levels.
- There are no special events or meetings in Kanban (or at least not as prescriptive as in the case of Scrum).

At the end of the day, the main difference between Kanban and Scrum lies in the way they are structured. While Kanban has its own ways of organizing tasks and teams, it doesn't prescribe specific techniques or roles, whereas Scrum does. This might be why Scrum is frequently seen as an easy-to-understand gateway into agile project management and why Kanban is frequently perceived as a "level up."

You are more than free to use a combination of the two methods. For instance, you might find the Scrum board is a bit more explanatory, and you might borrow it into Kanban. Likewise, you might find the Daily Scrum is a useful tool in managing your team. As such, you might want to bring it into your Kanban approach too.

There's no right or wrong as long as you abide by the general agile Principles and stick to the Manifesto. As you will see later on, the lesser-known (but equally valuable) project management methodologies we will present all show similarities to Kanban and Scrum but take things a little further from one point of view or another.

Benefits of Kanban

Benefits of Kanban for Software Teams

1 Continuous Delivery	6 Cycle Time	11 Focus on quality
2 No Estimations	7 Reduction of Waste	12 Pull Principle
3 Iterative Workflow	8 Frequent shipping, faster feedback	13 Never miss Blockers
4 Continuous Improvement	9 No Planning Overhead, less Meetings	14 Push Notifications with Integrations
5 Seamless Communication	10 Reduced PM Overhead	15 One-click Analytics

Like all agile project management methodologies, Kanban comes with a pretty generous list of benefits. How each project manager, team, and organization experience these benefits depends from one situation to another. However, some of the most popular advantages of incorporating Kanban in your company/ team include the following:

- Excellent level of flexibility. As mentioned in the previous section, Kanban is far less structured than Scrum. This can be confusing, sure, but it also opens the doors to a world full of opportunities. For once, you can maintain the traditional waterfall organization of the team and projects (and still benefit from all those agile practices). Even more, flexibility tends to be appreciated in modern workplaces not just because it helps

195

deliver better products but also at the team level. With Millennials now being the most numerous segments of the entire workforce (Emmons, 2019), this type of flexibility can earn you loyal, hard-working, talented people in your team.

- Continuous delivery. All agile methodologies focus on delivering on a continuous basis, yes. Kanban makes no exception and actually puts this at the core of the entire approach. As such, this can be considered one of the most preeminent advantages of using Kanban in your organization.

- Waste reduction. No matter how you look at it, waste is one of the major issues of the modern world. We waste paper, food, and time like there's no tomorrow, and while this definitely hurts our future as a race on this planet, it translates into far more immediate and palpable results in the world of business: money. Reducing waste means increased profits, and there's nothing higher management loves more than that!

- Productivity boost. Kanban is almost obsessive about keeping people busy with one task at a time. As such, the productivity of your team will increase. They will stop doing three things at once, and they will leave multitasking at the door when they enter the workplace. Soon enough, their productivity and efficiency will improve as well. Because, yes, studies show that nobody and *nothing* can actually multitask (Cherry, 2019)!

Obviously, all these benefits eventually transpose into actual business results. Happier and more productive teams deliver better products, they do it on time, they reduce waste in the process, and they tend to stick around for the next project as well. As such, the customer is more likely to be happier as well.

This is not to say that Kanban doesn't show any kind of disadvantages. On the contrary, you should be well aware of the fact that Kanban can lead to low-quality products if, for example, the Kanban board is outdated. Furthermore, Kanban tends to be less time oriented because it doesn't impose actual timeframes, and as such, this can lead to problems as well.

All in all, though, used right, and used in combination with other project management systems, Kanban can make your production process more fluent and more efficient across its verticals.

Kanban and VersionOne

As we were saying earlier, both Kanban and other agile project management methods can benefit from using the right tools. These days, agile project management tools are recognized at their true value, so much so that you won't see many agile teams without at least one of the major tools (which we have already discussed in the first installment of this book series).

VersionOne is one of the best tools to use when you want to implement Kanban because its very structure is built to be fully congruent to the basic Principles behind this specific agile approach. Although VersionOne can be used with pretty much every other agile method out there, it does provide a ready-made Kanban board that will help you implement the methodology in an easier, smoother way.

Some of the other features that will help you run your agile projects better include:

- the ability to connect distributed teams
- the ability to connect teams that use different agile methodologies (including Scrum and XP)
- the ability to automate project tracking
- the ability to collect data about the performance of your team and the efficiency of the entire project development process

We suggest you try VersionOne's free version and see how it works for you. Of course, you can use any of the other tools we have already mentioned in the first part of this book series, but if you are looking for something that is a little more Kanban inclined, this specific tool might be what you are looking for.

Kanban can be a real steppingstone for your team, and it can completely change not only how you work but how you perceive the entire concept of work itself. Although it might sound strict, Kanban is one of the more relaxed "entry agile" methodologies out there, one that will allow you to play around with team structures and agile concepts as you see fit for your specific needs, for your organization, and for your team's mental structure.

Implementing Kanban might not be 100% easy, but it definitely is one of the least headache-inducing agile methodologies to introduce to a team that has been working on a waterfall framework until now. Because it leaves plenty of room for documentation, charts, and other traditional project management tools, Kanban is a relatively good fit for larger organizations and corporations as well (and if you are ever in doubt, check out Toyota and how they use lean manufacturing and Kanban Principles in their work).

At the end of the day, no agile project management methodology can be a saving grace if you don't give it a real chance, and Kanban makes no exception. Study it, implement it wisely, and watch it bring your profitability, proficiency, efficiency, and overall productivity to a whole new level.

Our next chapter is dedicated to a more intense agile project management methodology: XP (Extreme Programming). The name shouldn't scare you, though. Same as Scrum and Kanban, XP abides to the same Principles of agile project management; it just takes it up a notch in terms of just *how* in-depth it goes with said Principles.

We definitely advise you to look into XP and the other project management methodologies we will describe further on in the book. They might not be as wildly popular as Scrum and Kanban, but they might be just what you are looking for!

Chapter 3: Extreme Programming (XP)

Extreme Programming (also known as "XP") was one of the first agile project management methods developed almost exclusively for the world of software development. Same as Scrum, XP came as a solution to problems that were specific to the 1990s (and which had been escalating since the 1980s, actually).

The fact that XP is more commonly used in software development environments makes it less known outside of the world of coding but no less valuable. The *extreme* nature of this methodology has somewhat washed out over the past two decades since its incipient stages, but the basic principles stay the same.

Indeed, XP seems to be a less popular choice in and outside the world of software development. Studies show that, in 2015 (Ropa, 2015), less than 1% of the respondents were using XP as their core agile methodology.

We believe it is still relevant, though. Even if you don't embrace it altogether and take the extreme path it suggests, this methodology can still provide you with valuable tools and practices, and many of them can be successfully incorporated into other agile approaches (including Scrum and Kanban, as it has been shown over the past chapter).

To understand Extreme Programming, you must first understand why there was a need for such a swift turn in practices. The origin of XP can be traced back to two main issues software development houses in the 1990s were facing:

- the need to adapt to an increasingly fast-paced world (consider the fact that the 1990s represented the rise to power of the dotcom bubble)

- the need to focus on object-oriented programming, rather than procedural programming

At its very core, XP is much more technical than other project management methods, and this might be one of the reasons it hasn't transcended the borders of the industry in which it was born (like Kanban and Scrum did, for example).

Extreme Programming relies on discipline even more than its agile "sisters." Its main goal is similar to those of every other agile methodology in the world: to deliver a better product in a shorter amount of time.

Planning/Feedback Loops

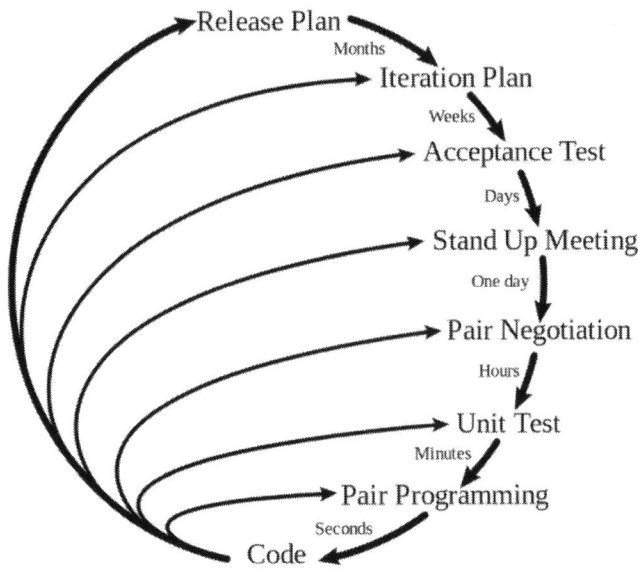

To reach this goal, XP relies on four main activities:

- Coding. In the Extreme Programming paradigm, this is the absolute key to developing quality products. Coding is King!

- Testing. While XP argues that codes are the most important element of the software development process, it also admits that testing *needs* to be done to ensure written codes are correct. There are three types of testing to be done: unit tests (to verify that each feature is functional), acceptance tests (to verify that the user requirements have been properly understood by the developers), and integration tests (to verify how different sections of a software connect to each other).

- Listening. Constant feedback and discussions with customers are encouraged so that developers understand the requirements. Moreover, developers are also encouraged to explain why some requirements may or may not be possible (and how they are or are not). Of course, since in most cases it is unlikely that the customer will be as technical as the developers, these explanations have to be provided in a business tone, rather than a technical one.

- Designing. In an ideal world, coding, testing, and listening should be *just enough*. However, XP recognizes that most software systems are complex enough to need proper design that connects the different features and makes the program easier to understand and use.

Aside from the core actions of Extreme Programming, it is also worth mentioning that this methodology has embraced the agile Principles and reshaped them in a form that is more congruent to the basics behind its specificities. As such, the values promoted by XP project management are:

- Communication. Developers have to constantly communicate among themselves, as well as with their customers.
- Simplicity. Everything should be kept as simple as possible, from the code itself to the processes behind the development.
- Feedback. As this is a basic tenet of agile project management, it hasn't been missed from this specific methodology either. Feedback is always encouraged and it should be followed at all times.
- Courage. In XP, courage is manifested in many ways. For once, the team must have the courage to program for today, rather

than tomorrow (which avoids long and winding design processes). Furthermore, they should know when to delete code and when they should leave it as it is. Last but not least, programmers should have the courage to accept that sometimes a problem might not be easy to solve but that persistence will eventually help.

- Respect. This value is manifested both at the level of the individual (team members should respect themselves and their work, and as such, they shouldn't make compromises) and at the level of the team (everyone should respect each other's work).

In essence, XP is not much different than other agile project management methods. What makes it different, however, are the specific practices it employs. We will dedicate the remainder of this chapter to exploring these specificities. As mentioned before, you can borrow all of them into your project management approach, or you can nitpick them and settle on those which prove to be valuable for your team's needs.

XP: Core Practices

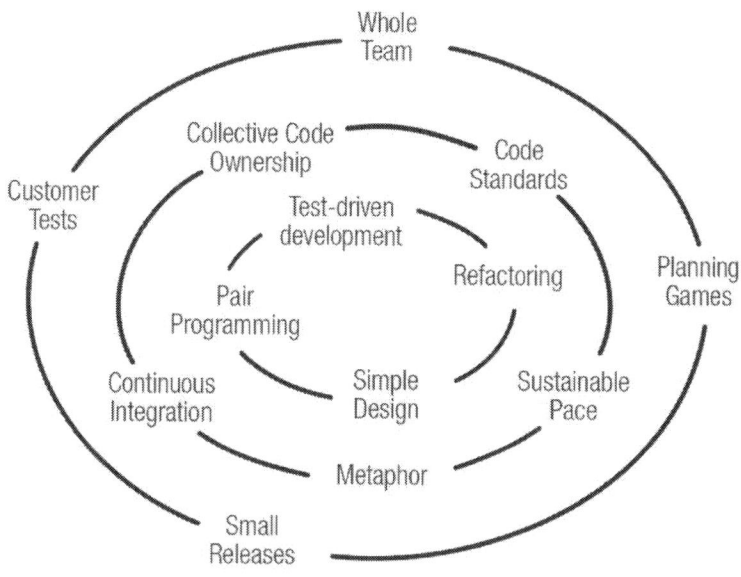

Planning Game

What is known in Extreme Programming as the "Planning Game" is nothing more than the specific way in which XP sees the planning of a project.

What is very specific about this planning method is the fact that the customer is always involved in the process. In short, the customer will bring forward all the information they have about the value of their requested product, while the development team will bring forward the information related to the cost of the product development.

The dynamics between the customer and the development team are meant to achieve one main goal: to reduce the time it takes to develop the product by removing intermediaries and helping the customer communicate directly to the programmers.

As you can see, this is quite a difference between XP and Scrum, as Scrum has at least two intermediaries between the customer and the development team (the Product Owner and the Scrum Master).

Although this is not exactly set in stone, most XP project Planning Games include four steps:

1. Create and select the story. If there is no story created, the customer must create one. If the stories have already been created, then the customer should simply pick one from the list.
2. Estimate the story. Once the customer has picked the story, the ball is in the developers' court. In other words, at this stage, the

programmers should estimate how much time it will take for the story to be delivered and how much it will cost.

3. Repeat the process with all the stories.

4. Prioritize the stories. Once all the stories have been estimated, they will be prioritized by the customer according to the requirements they want the final product to meet.

You will encounter similar procedures and games in other agile project management approaches as well (including Scrum, for example). The main difference lies in the involvement of the customer in the entire planning process (a feature XP puts a lot of emphasis on).

Small Releases

Splitting the project into smaller increments (stories) is very common in agile project management. The main purpose of doing this is twofold. On the one hand, it allows teams and project managers (where they are present) to better manage the workload. On the other hand, it allows teams and project managers to deliver better and faster.

Small releases are an integral part of Extreme Programming as well and, same as with other agile methodologies, their main goal is to allow for better control in terms of timely delivery, quality, and adaptability to change.

In XP, small releases translate into actual mini-product releases that are made available to the customer. Once feedback and data are gathered on what works and what doesn't, the product will be improved throughout the following iterations until it is ready to be published in its full format.

To make sure your small releases achieve their main goal, you should keep in mind to:

- Be ready. Just because this is an agile approach, it doesn't mean that you shouldn't plan ahead. In XP, you are required to know what each iteration/ story will deal with, as well as how long each iteration will take to be completed.
- Communicate. You should *always* make sure to communicate with the customer, as well as among yourselves, as a team. All the features should be discussed, both in terms of how-to time and budget them and in terms of how to actually code and execute them.
- Feedback implementation. As we have also mentioned before, there is no point in asking for feedback if you don't take action accordingly. Continuous improvement is crucial in all agile project management methodologies!
- Experiment. You shouldn't be afraid to experiment with new methods, new codes, and new solutions. At the very worst, they will not work. At the very best, you might have just found a new way of solving a problem that is most likely older than your project.

Small releases can be extremely helpful when you want to make sure you have granular control over what happens in the project and when you want to make sure the end product is fully satisfactory for the end user.

Customer Acceptance Tests

We have already touched upon the concept, but we would also like to expand on it a little more, since it's a practice quite specific to Extreme Programming (one that can be incorporated in other agile approaches too).

Customer acceptance tests are quick tests (frequently automated tests) that involve the customer in a measure that is equal to their involvement in the Planning Game. Basically, what the customer will do is translate each story into an acceptance test. When the story is delivered, the customer will specify different scenarios to see if the product functions as they wanted it to function. If it doesn't, the product will go back into development, and the feedback will be incorporated in the new iteration.

Before the product reaches the customer's hands, it is quite important to ensure that it has been properly run through the Quality Assurance team's tests. This also means that programmers and QA engineers should have a pretty close relationship and that they should communicate with each other quite efficiently (especially given that there is no project manager here to mediate the relationship between the two).

No user story should be considered as "Done" until it has been approved by the customer. One story can also run one or more customer acceptance tests, depending on the complexity of the product and how good it is when it is released the first time.

Simple Design

In agile, everything should be maintained at a very high level of simplicity. Everything should be easy to work with, easy to implement, and easy to use when it comes to the end user.

Simple design is one of the core values of Extreme Programming because it allows the very architecture and presentation of the product to stay clean and tidy. At the same time, it is essential to mention that "simple design" is not one and the same with "simplistic design."

You want complex programs to be easy to use. You don't want to ultrasimplify the complexities of your program. To ensure this happens, keep in mind to:

- Do your research and be aware of any potential flaws in your design.
- Code simply; do not overcomplicate it unless it is absolutely necessary.
- Don't be afraid to postpone decisions to a moment where you can have full insight into the data your business is collecting.
- Document all the design choices and changes in your backlog, so that you can go back to them as a reference for the future.

Simple design can help you make the development process easier for your customers, and it can help you deliver the kind of product they can easily use. While your first story might not be exactly what the customer is asking for, these feedback cycles are actually helpful because they will help you understand everything better.

Don't look at simple design as an impediment on your way to success. It is there to help!

Pair Programming

Pair Programming is a concept we have already introduced in our previous chapter, one that has been borrowed not only by Kanban (as we have shown) but by other agile methodologies as well.

Pair Programming brings two developers together. One is the Driver (the person who writes the code) and the other one is the Observer (the person sitting in the back, verifying the code written by the Driver). Ideally, both the Driver and the Observer will change places and one will become the other, so that they can both have the same amount of time in front of the keyboard.

Moreover, it is recommended that both the Driver and the Observer have the same amount of experience under their belts. This way, neither one of them will feel left behind or "pushed" in any way.

The Pair Programming practice relies on the four-eyes principle applied in multiple industries. For instance, in book publishing and translations, it is believed that two pairs of eyes are more likely to spot a variety of mistakes (as opposed to having just one pair of eyes check the work). Furthermore, in business, you will frequently notice that documents have to be signed by two people (e.g., the CEO and the CFO). This too is a permutation of the four-eye principle.

In XP and the other methodologies that have employed Pair Programming, the four-eye principle relies on the idea that two people are more likely to write clean code. Both the Driver and the Observer have to work together by coding and testing the features they are building, and, in the process, both of them are more likely to spot mistakes along the way.

Pair Programming brings with it a long list of benefits, such as boosting collaboration and communication, improving the overall quality of the code, and delivering in a shorter amount of time.

One of the main drawbacks associated with Pair Programming is that simply having the hard skills (the coding knowledge) is frequently not enough for this to work. In order for two people to work as closely as this, they need to also possess soft skills (like the ability to communicate properly and a general ability to work as a team, for example). Without the soft skills, it is easy to see how this practice can very quickly turn into a bad joke at the very best or a complete disaster at the very worst.

Test-Driven Development

The concept of Test-Driven Development is a very good fit for the world of agile in general, and even more so when it comes to Extreme Programming, where things tend to be even more on fast forward.

Test-Driven Development is a type of software development process that focuses on a very specific, very short development cycle:

- User requirements are turned into test cases.

- The tests are run for verification.
- The coding is done.
- The tests are applied on the code.
- The code is cleaned during the refactoring phase.
- Everything is repeated until the final product is delivered.

Test-Driven Development shows one major benefit: it helps programmers stick to cleaner codes, as opposed to getting lost in intricacies of any kind. Principles such as *Keep It Simple, Stupid* and *You Aren't Gonna Need It* are commonly applied to Test-Driven Development precisely because it allows the developers to focus on clean code that is well designed and simple to use, rather than overdoing it and creating a code that will be more difficult to change or clean up.

There are some practices you should avoid in Test-Driven Development as well:

- Your test cases should not depend on the previously executed test case systems. Instead, you should start a unit test from a preconfigured state at all times. Likewise, interdependent tests should not be run either, as they can lead to false negatives and fail the entire process.
- You shouldn't go overboard with your test cases either. For instance, trying to create a test that will inspect everything under the Sun means that you will make your entire test more prone to mistakes, and that defeats its purpose.
- Your tests shouldn't be slow running either. Since this is a very fast-paced environment, both your coding and your testing have to move at the same speed as the project. Of course, this is not

to say that you should sacrifice the quality of the code, but you shouldn't linger too much on any of the steps involved in the development process.

Test-Driven Development may not be all about advantages, for certain. But in some cases, using this technique might actually make the difference between a project delivered on time and one that is late.

Refactoring

This is one of those concepts that sounds much more difficult than what it actually is. Commonly used in XP and other agile methodologies, Refactoring is a practice that focuses on simplifying the code to the point where changes and bug fixes can be very easily implemented.

At the same time, it is very important to mention that Refactoring *does not* mean that you should sacrifice the functionality of the code in any way.

To ensure the efficiency of the Refactoring process, it is of the utmost importance to be constant in practicing it. In other words, Refactoring should be a regular affair. You can do it daily or weekly, but it needs to be done with regularity, every single time. Moreover, it should follow the same simple process every time as well. For the team, Refactoring should become like a second nature, a routine they implement into their work in the same manner they grab a cup of coffee every morning when they get to the office.

The Refactoring process can be adapted according to needs, but most often it will follow the same pattern:

- Find or create a test for the part of the code you want to refactor. This will help you make sure you do not ruin the integrity of the code in the Refactoring process.
- Run code simplification and improvement as you see fit.
- Apply the test on the unit you have modified.
- Repeat until you have Refactored all the units that need to be changed.

The Refactoring practice is simple and it can be a real lifesaver. Say, for example, that your customer wants to change something about the product when it has already been developed. The simpler the code behind it is, the easier it will be for the team to dig in and apply the needed change.

It's a win-win situation for everyone: the customer is happy with the speed of execution and with the overall quality of the product, while the team is happy that they don't have to navigate ridiculously complex codes to determine exactly where the change should occur (and how).

Continuous Integration

Continuous integration is not a concept that is exclusive to Extreme Programming in general. Rather, it is a concept that keeps popping up in the entire world of agile project management. However, it is more common to see this practice applied in XP and the other more extreme agile methodologies precisely because it relies a lot on automation and on fast-speed development processes, which makes it more suitable for these specific agile approaches.

Continuous integration is a practice that works especially well in agile project management precisely because agile focuses on developing products in an incremental way. With the help of continuous integration, the different stages of development are more easily integrated into the whole without too much time invested in the integration process and without the potential errors that might ensue.

In short, continuous integration refers to keeping the code in a central repository where developers and testers can easily access it. When the system is ready, the code can be tested right away, using the aforementioned repository. This allows the team to run tests as they move along with the code, thus minimizing the time needed to deliver working software.

These days, continuous integration is even easier to perform than what it used to be like back when Extreme Programming was first shaping up. This is mostly due to automated systems that allow programmers to easily move along with the code and test it at the same time. Machine learning has made everything easier, allowing for better productivity and efficiency in software development (and not only these areas!).

Collective Code Ownership

As we were mentioning before, Extreme Programming is an agile methodology that is quite extreme in the level of flexibility it allows. The development of an XP project is *so* flexible that everyone is encouraged to constantly bring improvements to any part of the project that they might want to.

The code is collectively owned by everyone in the team or, in the case of open-source projects, by everyone who wants to chip in and simply make the code better from one point of view or another.

The XP team is completely decentralized and destructured, and this might be a concept pretty difficult to grasp if you come from a very traditional environment. In fact, it can be pretty difficult to grasp wherever you may come from, including the more structured agile methodologies out there (such as Scrum and Kanban, for example).

In an XP project, everyone is responsible for the entirety of the code. This means that the team doesn't rely on a chief architect to have all the good answers, all the time. Instead, the team relies on its entirety to make the software a working product.

Extreme Programming understands, perhaps better than many other project management methodologies, that nobody is perfect, not even a super-experienced chief architect. Instead of placing the responsibility for the final product on one or two people, XP shares the load. As such, the code is collectively owned, and everyone is invited (and encouraged) to participate in it.

This might sound like chaos, but years and years of experience have proven that, yes, this is a more reliable way of treating software development. With continuous integration being such a massive part of

Extreme Programming projects, the amends made by the team to any part of the code are hardly even noticed. Instead, everything simply flows smoothly.

Collective code ownership also means that you have to place complete trust in your team and their skills. Without this trust, it's easy to see how you might become suspicious of your team members and how you might not be able to see their improvements as actually beneficial.

On the upside, this level of trust in a team can also empower them and make them more responsible at an individual level. It might be surprising, but most of the times, people don't disappoint when so much trust and responsibility is placed in their hands. Rather, they constantly double- and triple-check all of their work to ensure that their codes and their amends will not have to be cleaned up by the rest of the team.

Coding Standards

One of the major downfalls of practices like the Collective Code Ownership is related to the fact that, well, different people can see a feature developed in different ways.

Programming is very much like every other language: you can use it as you see fit and still get the message across at a basic level. As such, you need standards to ensure that people in the same context will *speak* the same language, not just in terms of basic linguistics but also in terms of tone, style, and voice.

The same principle applies to XP projects as well. Coding standards have to be set in place to ensure the consistency of the code across the entire project. Otherwise, you might soon learn that there is a major issue with so many people being responsible for the entirety of the code: they think differently, and they might not be on the same page in terms of how they organize their code, how they relay the information to the computer, or even how they understand the concept of "Done."

Although there is no prescription for this, most often, there are three main categories of coding standards:

- what is mandatory (the rules that have to be followed at all times)
- what is considered to be a good practice
- what is considered to be a recommendation (tips that should be followed, but they are not mandatory, and the developer should use their better judgment in deciding whether or not to use them).

Furthermore, there are different types of coding standards as well:

- formatting (such as, for example, how white spaces are used)
- code structure (such as, for example, project layout, classes, resources, and so on)
- naming conventions (how the team names their classes, methods, and so on)
- error handling (how objects in the code should handle errors, reporting, and logging, for example)
- comments (how to explain the logic of the code)

Coding standards are quite important when you want to minimize the risk of a messy code (and thus, a code that is more likely to have bugs and less likely to be easily adapted to the changing requirements of the customer).

Metaphor

The System Metaphor (or more simply put, the *Metaphor*) is, at its very core, nothing but the communication that backs up the development of a product. The Metaphor is a truly essential part of an XP project (and it is encountered, in different forms, in other agile project management methodologies as well). At the same time, the System Metaphor is also very poorly understood (and commonly forgotten altogether).

Basically, what the Metaphor does is communicate with the customer and the stakeholders in their own language. It is a translation of the technicalities and the jargon that are used within the team for people who are involved in the project but aren't exactly "technical."

There are two levels at which the Metaphor happens:

- A unified story: a story that hovers over the entirety of the project and must be used by all those involved, at all times.
- A shared vision among the different participants in the project. No matter where they come from, everyone involved in this project will have a shared vision (coordinated by the unified story, of course).

The way you use Metaphor in your project can actually make the difference between a good understanding of the product requirements and a bad one, between customers who do understand your processes and customers who don't, and, in the end, between products that satisfy the needs of the clients and products that don't.

Although it might be a less technical part of the XP project, the Metaphor is as important as any other practice associated with this agile methodology. Use it wisely and it will help you deliver better products!

Sustainable Pace

Sustainability and scalability have been frequently touted as the two major downfalls of agile project management in general.

The sustainability of XP has been especially problematic and argued over the decades, and the debate that ensued lies at the basis of why Extreme Programming isn't as popular these days as it used to be (or at least not in its purest form).

Due to the fast-paced environment of the software development world in general, many XP teams used to feel the need to go at a speed that was much higher than the average. As a result, burnout and mistakes happened as well, proving that going well overboard with what you can do is never a long-term plan.

Contrary to what many people would be tempted to believe, XP does not advocate for a lack of sustainability. On the contrary, proper

Extreme Programming should be done at a pace that is sustainable in the long run.

There are multiple reasons for this. The first one is related to the fact that it is simply impossible to expect a very high speed of delivery and a very high quality of product at the same time or at least not long term. It might happen for one sprint or two, but it cannot be a sustainable strategy for the duration of an entire project, much less a project that will be finished in a matter of *years*.

Furthermore, Extreme Programming is highly reliant on team members' abilities to actually cooperate and communicate with each other. Unfortunately, that rarely happens in teams that hardly know each other due to high job mobility. Unfortunately, as well, job mobility and unstable teams are two of the consequences of burnout.

As such, it is highly important to make sure you move at a pace your team feels comfortable with. It may not lead to super-fast delivery times, but it will maintain the integrity of the code, and maybe even more importantly, it will maintain the integrity of the team.

Extreme Programming has been dubbed as one of the least scalable agile methods, and we are sorry to say it, but we tend to agree with this statement. Because it relies so much on the cohesion of the team, XP cannot be scaled to very large teams and corporations.

What *can* be done, however, is integrating XP-specific practices (such as the Paired Programming one) into other agile and/or traditional methodologies. Doing this will allow you to get the best of the two worlds: the high speed and high quality that XP projects are frequently associated with *and* the long-term sustainability of applying such methodologies to projects and organizations that go beyond your average XP team size.

At the end of the day, it can be quite easy to understand why XP has decreased in popularity. Instead of remaining closed within the limitations of each methodology, agile has proven, time and again, that flexibility is its main tenet, and, as such, it has grouped together practices pertaining to different methodologies to create the perfect environment for each situation.

Scrum, Kanban, XP—none of the major agile project management methodologies are used in their pure forms, most of the time. Instead, companies and project managers rely on a combination of practices that are fit for their teams and their specific situations. We encourage you to test out Extreme Programming practices too. They might prove more than useful!

Chapter 4: The Crystal Method

While you may have heard about Scrum, Kanban, and even Extreme Programming, the Crystal Method might be more of a mystery to you if you are new to the agile project management world.

Indeed, the mainstream agile methodologies have migrated past the borders of their agile world, and they are now common even in companies that deem themselves as adhering to a traditional waterfall approach.

When it comes to the less popular methods, however, things get a little more complicated. Methods like the Crystal one is incredibly powerful and useful for a wide range of situations, but they tend to be less prescriptive and thus less adaptable to teams that are new to agile. As a result, the Crystal Method has not yet made the leap into the mainstream agile project management framework.

We want this book to be a comprehensive overview of the main agile methodologies employed by worldwide organizations. As such, we have decided that we will include the less popular, but still extremely valuable, methodologies here as well. Crystal falls in this category because, as you will see later on, it can provide you with a lot of benefits.

We do not claim that this chapter covers everything Crystal is about. As mentioned in the introduction of this book, we do not claim that the entire book encompasses everything all these agile methodologies are. Instead, our purpose with this specific chapter is to introduce you to the Crystal Method and allow you to make your own choices. You might be surprised to find that Crystal makes a lot of sense, and you might decide that it is the right direction for your team and organization too.

Even if you don't choose Crystal to guide you through the intricacies of the agile world, we still believe there is value in learning about it. The more you know, the more in depth you can go with your agile approach and create a "program" that allows you and your team to perform at your maximum capacity!

What Is the Crystal Method?

This is a lesser-known fact, but the Crystal Method (or the Crystal *Methods*, as the approach is frequently referred to) was born in one of the biggest and most resilient IT companies in the world: IBM.

It sounds odd to think that one of the least known agile methodologies out there was born in one of the most famous IT businesses, but the low-prescriptive, very lightweight nature of Crystal explains why it isn't one of the more popular choices for companies these days.

In addition to this, Crystal is one of the first agile methodologies ever developed, at the beginning of the 1990s. Come to think of it, there was a one-decade gap between the moment Crystal was developed and the first official Agile Principles were laid down in Utah, so it makes quite a lot of sense that the Crystal collection of practices has been left behind in the fog of time.

This doesn't make Crystal any less valuable. On the contrary, we believe that it can be a very good approach, especially for teams that are already at least somewhat used to agile project management.

Sometimes, it takes going way back to the roots to discover the true essence of something that has evolved over the years, and agile project management is no exception to this general statement. The Crystal Method might be exactly what you need to rediscover the true roots of agile and why it was such a groundbreaking, earth-shattering framework when it first started to trespass the closed world of high-end software project management.

The Crystal Method is very simple in essence, and yet, applying it with no limitations whatsoever can be quite tricky, precisely because it leaves the door open to a world of opportunities in terms of the specific practices you may or may not want to employ with your team.

The complications of Crystal arise when you try to understand the fact that its value lies in the lack of prescriptiveness and the ways in which it can be adapted to teams that are wildly different in nature.

In other words, Crystal is only as hard as you make it. Beyond the basics (which we will relay in the following sections of this chapter), Crystal is nothing more than one of the incipient agile methodologies. It relies on the same basic Principles; it's just that they are less polished, less prescriptive, and less limited at the same time too.

If we had to summarize the Crystal Method, we would simply say that it is a collection of primordial agile practices adapted to the specificities of various types of teams. Although incipient and apparently dated, these practices continue to make sense in today's environment, particularly as more and more companies are looking to scale up the agile methodologies they have been using to date.

How Does Crystal Operate

When he first analyzed the dynamics of teams at IBM, Allistair Cockburn noticed that there are two main tenets that make them more efficient:

- the ability to streamline and optimize their own processes according to the workload
- the adaptability they have when it comes to the uniqueness and specificities of each project

Later on, these two concepts became the foundation of the Crystal Method in a way similar to how the Agile Manifesto rules over the entire agile framework.

While he was developing the Crystal Method, Cockburn also noticed that there should be a clear differentiator between methodology, technique, and policy. This helped IBM understand where Crystal begins and where it ends, and it continued to help organizations everywhere delineate between what their project management approach is and what other regulations of their businesses are.

In short:

- A methodology is a set of elements (such as practices and tools, for example).

- A technique is related to skill areas.
- A policy is related to what *must* be done in an organization.

Furthermore, Cockburn also stated that the new set of methods he was developing would be focused on six main areas:

- the people
- the interactions between the people
- the sense of community people gets at work
- the skills people have
- the talents people have
- better and more efficient communication between the people

As you can see, the Crystal Method is a clearly human-centric view on project management. By far and large, what Cockburn developed back at the beginning of the 1990s might have set the foundation for the first point of the Agile Manifesto later on, in 2001.

Same as the Agile Manifesto and its adjacent Principles, Crystal says that processes are never more important than the people. In fact, according to Cockburn, people should come first in all circumstances, and processes should come second. The *team*, as Cockburn puts it, is the core of the project because they are the ones holding the skills and the talent, and as long as these are abundant, processes follow after.

Additionally, Cockburn also identified four main behaviors people have in relation to work:

- People are meant to communicate and crave for proper communication. It is worth mentioning that this is a concept all agile methodologies support. But it is equally important to mention that this concept was perceived differently back in the 1990s and the beginning of the 2000s, when online communication had not yet reached the level of accuracy and real-timeness of today's technology.

- One of the main issues people have is being consistent over the course of a longer period of time. As such, the project management approach should be built in a way that keeps the team continuously engaged and supports software development in a sustainable way.

- People are not constant either. They have good days and bad days, they may or may not perform at their best in one place or another, and they will always be the same. As such, the processes behind the project management approach should adapt to this type of change as well.

- People are inherently good. They want to be good citizens, they want to take care of their peers, and they want to do a good job. If you enable them to be their best, they will be, and as such, they will perform better on their day-to-day duties at work.

Although this might sound odd coming from one of the largest corporations in the world, the Crystal Method is one of the easiest ones to adopt in terms of the rules it brings forward precisely because it has designed mini-frameworks for a variety of situations.

More specifically, Cockburn split the entire Crystal Method into several categories, organized according to five major criteria:

- the size of the team/ project
- comfort
- discretionary money

- essential money
- life

The first criterion is pretty straightforward: the size of the team influences how it should be managed.

The other four, however, are related to the impact the system could have on the mentioned verticals if it doesn't work:

- how it will impact comfort
- how it will impact the disposable income of the user
- how it will impact the essential income of the user
- how it will impact the life of the user

According to these five criteria, Cockburn has categorized the Crystal family of Methods into five color-coded groups, as follows:

1. Clear: for teams of up to 6-8 people, low to no impact across all verticals
2. Yellow: for teams of up to 20 people, low to medium impact across all verticals
3. Orange: for teams of up to 40 people, medium to high impact across all verticals
4. Red: for teams of up to 80 people, high impact across all verticals
5. Maroon: for teams of up to 200 people, very high impact across all verticals

Aside from these five basic categories, you might also encounter adjacent Crystal Methods used in specific contexts:

- Crystal Orange Web (used for web products)

- Crystal Sapphire and Crystal Diamond (used for large-scaled projects that can have a dangerous impact on human life, like aircraft software, for example)

It is also essential to note that there is quite a lot of flexibility when it comes to how each category applies to your team and project. However, should the team grow over time, it is recommended to upgrade to the next color-coded category, rather than continue to apply the same methods and try to scale them up.

This might all sound very prescriptive, but it is quite the opposite in most respects. At the same time, keep in mind that this methodology was not born in a small software development house. It was born in a company that was already a mammoth corporation by the time Crystal came into play. As such, you should see the lack of prescription through the prism of a company that was very well-grounded in traditional approaches too.

One of the areas in which this is more apparent is the way in which Crystal Methods assign specific roles to team members and stakeholders. While you will encounter a pretty deep level of role-defining in Scrum as well, the very nature and the names of the roles assigned in a Crystal project will most likely be more familiar to those coming from waterfall approaches.

Some of the roles you will encounter in a Crystal project include the following:

- Project Sponsor (customer or internal stakeholder)
- Senior Designer/Programmer (the equivalent of team leads)

- Designers/Programmers (Business Class Designers, Programmers, Documenters, etc.)
- Architect
- Requirements Gatherer
- Coordinator
- Business Expert
- Project Manager
- Design Mentor
- Lead Design

... And so on.

We will not dwell too much on the specific differences between the different roles assigned in a Crystal project. For the most part, they are quite self-explanatory and quite similar to traditional roles you will encounter in many other software programming projects.

Furthermore, we will not dwell too much on the differences between the different Crystal family members either. Each of the colors is assigned to a type of project. Different specific methods are meant to be employed according to the criteria that were already tackled earlier in this section.

If you want to take a specific Crystal approach, we highly encourage you to go in depth with your knowledge on that specific family member. Same as in the case of XP, even if you don't end up using that particular approach, you might still find practices your team is congruent with.

Crystal Method Characteristics

The basic characteristics of the Crystal family of Methods might be easy to understand from the introduction we have made in the previous section. However, for the purpose of clarity, we feel the need to also relay the specific characteristics that bring all Crystal Methods together.

Things are quite simple in this sector. There are three main characteristics the Crystal Methods have in common:

1. They are all human-powered. As it was shown in the previous section, manpower lies at the very core of the Crystal methodology. People are the blood and veins running through projects and making them come to life, and, as such, all Crystal Methods (regardless of the color they were assigned) put massive emphasis on the importance of the human resources.

2. They are ultralightweight. All agile methodologies are lightweight, but methods like Crystal take this up a notch and become *ultra*lightweight. What this means is not that team members are allowed to do whatever they please, but that documentation and reporting are less of a point of focus in Crystal than they are in other methods. This is also why implementing Crystal from a waterfall standpoint might not be as easy as it is generally believed by those who go through its basic principles and M.O.

3. They are adaptive. Like all agile methods, Crystal methods are adaptive. They embrace change as a natural part of the process, and they embrace the differences between the different teams

as an equally natural element. As such, Crystal Methods are adaptive, flexible, and interchangeable down to their very core.

That's it. These are the characteristics that bring all the Crystal Methods together and tie them to the larger world of agile project management.

Properties of the Crystal Method

Aside from the basic characteristics shown in the previous section, the Crystal Method has a very stable set of properties as well. As you will see, these properties are similar to the Agile Principles. So, for the most part, they will most likely not surprise you.

The seven main properties of the Crystal Method are as follows:

1. Frequent delivery. Same as all the other agile projects, Crystal projects should be delivered in smaller increments, rather than larger chunks (or entire products at the end of the project). This will help you ensure that you don't put a lot of money and energy into a product that will not be well received (by the customer or by the end market).

2. Reflective improvement. Constant and continuous feedback lies at the foundation of agile in general. So, it makes all the sense in the world that Crystal would adopt this principle as well. The more feedback you get and the more you reflect on how to improve your product, the better the end result will be.

3. Osmotic communication. This might sound very fancy, but it refers to a very high level of communication among team members. For smooth projects, you need to constantly encourage people to talk to each other and communicate their issues, opinions, and suggestions.

4. Personal safety. As mentioned before, Crystal is one of the most human-centric approaches you will ever encounter. As such, it puts a lot of emphasis on personal safety as well. This is not so

much related to personal safety in the sense of ensuring a lack of life-threatening conditions at work (which goes without saying, right?), but in the sense that people should be encouraged to speak up. Yes, this means they should speak up even when they don't agree with the majority.

5. Focus. This property is twofold. On the one hand, it speaks about each team member's ability to focus on one task at hand and how to deliver it better. On the other hand, it speaks about the focus of the entire project, defining clear objectives and goals and ensuring the entire team is on the same page in all respects.

6. Easy access to expert users. In Crystal projects, you are encouraged to always get feedback from the end users. Focus groups and surveys might help with this, but do keep in mind that you might need interdepartmental cooperation for this.

7. Technical environment focused on test automation, configuration management, and frequent integration. The more automated your processes are and the better your management and tracking tools are, the more likely it is that your team's skills and talents will be focused in the right direction. As such, the final product will be better from all points of view.

Why Is the Crystal Method Useful?

All agile project management methodologies are useful in their own way. The Crystal Method makes no exception in this area either.

Overall, the Crystal Method is considered to be useful for the following reasons:

- It's fairly easy to implement, both due to the high level of categorization it includes and to the high level of flexibility it allows.
- It's easy to implement its specific practices with other agile and nonagile methodologies.
- It focuses on people and communication, and in a world that seems continuously disconnected from its human nature, this is a *big* advantage.
- It allows continuous integration and incremental deliveries, which helps deliver better products in the end.
- The processes are configurable, and you don't have to worry about documentation too much.
- It actively involves the end user, and this means that you will get better, more accurate feedback that suits the actual market.

The lack of specific guidelines can be quite confusing when it comes to the Crystal Method, but it is more than worth mentioning that Crystal was not meant to become a policy or a set of actual techniques from the very beginning (this is specifically why Cockburn made the distinction between the three terms).

At the end of the day, Crystal can provide you with a mindset at the very least and a pretty good ultralightweight methodology at the very most. It is up to you and your team what you choose to do with the basic information you have been given!

Chapter 5: Feature-Driven Development (FDD)

Together with Crystal (which we have discussed throughout the last chapter) and the Dynamic System Development Method, the Feature-Driven Development approach (FDD) is less popular as a methodology per se.

This is largely due to the fact that FDD tends to be associated and annexed to some of the more popular methods, instead of functioning as a standalone methodology in its own right.

This is not to say that Feature-Driven Development is not important or that it cannot function on its own, on the contrary, actually. It can and you can definitely try it out.

Rather than debating whether or not FDD is popular or relevant, or whether or not you should actually implement it as a pure methodology, we want this chapter to be an informative one that will set things on a clearer path for you and help you understand not only what FDD is and where it stands in the world of agile but also how you can incorporate it in a unique and personalized approach to agile project management in general.

At its very core, Feature-Driven Development is not much different than many other agile project management methodologies. The one quality that differentiates it from the rest of the agile methods is the fact that, as the name suggests, it focuses on making progress on each feature.

Basically, the importance of the "story" delivery you know from Scrum is transferred to the importance of each feature. Sometimes, a story might coincide with a feature, but this is not a must.

Developed at the end of the 1990s for a project that aimed to deliver a short time-spanned project for the banking industry in Singapore, FDD is nowadays used quite widely. Many people automatically incorporate it in their Scrum or Kanban approaches, but it is important to note that just because this is common, it doesn't mean FDD and Scrum are one and the same (or that FDD and Kanban are, for that matter).

Feature-Driven Development is one of the more prescriptive agile methodologies out there in the sense that it works based on a clearly defined life cycle, and it assigns clear roles among the different team members.

The FDD life cycle is defined by five main stages at which the product is developed:

1. Developing the overall model. At this stage, the team gets familiarized with the high-level scope of the entire project and system, as well as the context it comes from. Further on, the team will split into smaller groups, and each area of the system will be modeled then presented for peer review. The model(s) that are deemed a better fit will be selected to act as a domain area model. In time, all the domain area models will be merged into the larger model.

2. Building a feature list. Once the knowledge has been collected throughout the first stage of the development, all the information will be used to identify the list of features. The domain area will be split into multiple subject areas, according

to the functionality of the features representative of the area. It is important to note that each feature should be identified through the prism of the value it provides the user with.

3. Planning by feature. Once you have the entire list of features at hand, you will be able to start planning the development per se. During this process, you will assign the feature sets as classes to the programmers in your team.

4. Designing by feature. Each feature of the product comes with a design package that will be handled during the fourth stage of development. This will be handled by a chief programmer in a team, who will initially select a set of features to be developed over the course of two weeks. Together with the class owners associated with these features, the chief programmer will create sequence diagrams for each of the selected features and use them to refine the model.

5. Building by feature. Once the prework has been done and once the design inspection is ready, it is time for the team to move on to the actual programming, then test and inspect the code to ensure the feature is complete. Once that happens, it can be incorporated into the system.

As for the roles encountered in FDD, they are quite clear, and they may include the following:

- Domain Manager
- Language Guru
- Build Engineer
- Release Manager
- System Administrator

- Tester
- Technical Writer

In some respects, FDD and XP are quite similar to each other. It is very important to note, however, that the major difference between the two comes with the introduction of a "class ownership" concept. As mentioned in the third chapter of this book, collective ownership is one of the particularities in Extreme Programming. In Feature-Driven Development, however, this ownership is transferred to the class owner who becomes responsible for its functionality.

The most notable tip of information you should keep in mind when it comes to how FDD functions is that instead of focusing on more or less random chunks of the project, it approaches development on a feature-by-feature basis.

Same as in the case of the other agile project management methodologies, Feature-Driven Development is associated with a series of best practices too. We will tackle them in the remainder of this chapter, one by one.

Domain Object Modeling

Domain Object Modeling is one of the most important best practices associated with FDD. This practice basically consists of both exploring and explaining the domain of the problem at hand. Once the domain object model is generated, the team will have an overall framework to use when adding features, one by one, as they are developed.

Developing by Feature

As mentioned before, this is one of the main tenets of Feature-Driven Development (as the name of this methodology suggests a well). There is one important rule associated with the concept of developing by feature: if a function is too complex to be developed and implemented in two weeks, it will be split into smaller functions until each of the resulting subproblems is small enough to be considered a feature in the full sense of the word.

This allows for better control of the changes that might appear along the road, be they related to changes in product requirements or changes related to bugs and/or lack of functionality.

Component/Class Ownership

This best practice was also touched upon in the introductory section of this chapter, but it is quite important to keep it in mind, so we will mention it here as well. What class ownership means is that different pieces or groups of pieces of code (called "classes") are assigned to specific owners. Each of these owners is responsible for the code on multiple levels:

- consistency
- performance
- conceptual integrity

Feature Teams

Same as most agile methodologies, Feature-Driven Development prefers working with small teams. As such, feature teams will be assigned for the development of each feature.

A feature team is a small team formed dynamically to develop a small activity. Together, they work for each design decision, and they evaluate their options before they choose a particular one.

As you can see, there's quite a lot of trust placed in each and every member of these teams, as well as in how they can work together and cooperate for the success of the entire project.

Inspections

According to FDD's best practices, you should run regular inspections of the design and code, so that you can detect any defects in due time (and so that you can apply the changes necessary to improving these defects).

Configuration Management

Configuration management is quite important, especially when new team members join in or when you are adopting FDD for the first time. However, it is a best practice that should be maintained regardless of where in your journey to Feature-Driven Development you may be.

Basically, configuration management will allow you to identify the source code for everything (all the features) that have been developed to date. This will also allow you to keep track of the class changes while feature teams work on their enhancement.

Regular Builds

The concept of "regular builds" is similar to that of "continuous delivery" in the general Agile Principles. What this concept refers to is ensuring that your system can demonstrate, at any time, that integration errors have been fixed early on in the process. This allows the customer to maintain trust in you and your team, and it allows you to actually keep track of all the issues, whether or not they are repeat offenders, and whether or not they have been successfully fixed.

Visibility of Progress and Results

This concept is quite similar to the Scrum and/or Kanban board in the sense that the progress and the results of your team's efforts should be visible at all times. FDD may not employ an actual board for this, but the FDD project manager needs to ensure that all reports are frequent, accurate, and appropriate both from the internal point of view (team, internal stakeholders, higher management, etc.) and from the external point of view (customer).

Feature-Driven Development is, as mentioned in the beginning of this chapter, a less common methodology on its own. However, it is quite essential to remember that it can be easily integrated with other agile methodologies.

In fact, the first two stages of the FDD development cycle are almost entirely congruent to the initial envisioning model of Agile Model Driven Development, showing that FDD belongs in the agile family just as much as Scrum, Kanban, or Extreme Programming.

Indeed, there is less focus on people in FDD (as compared to Scrum, Kanban, XP, or Crystal, for example), as the main center of this approach lies on the actual feature and how the *people* around it help with its harmonious development.

However, this doesn't make Feature-Driven Development any less agile in nature. At the end of the day, FDD abides to the general Agile Principles just as much as any other methodology in the book.

It is easy to dismiss the lesser popular agile methodologies out there, including FDD, especially since there seems to be less information and a smaller number of tools that are designed specifically for this approach.

However, what you do have to keep in mind is that methods like these are more of a supporting system to the more popular methodologies that focus on *mindset* more than *specific approach*.

FDD can work in combination with most of the project management methodologies we have described so far, with the exception of Extreme Programming (where the concept of class ownership and that of collective ownership will clash).

Same as in the case of Crystal, we thoroughly encourage you to learn more about FDD as well. Aside from the five main stages we have described here, each of them is associated with specific substeps that will allow your entire plan to be more structured. For this reason, FDD is one of the agile methodologies that seem to be more of a better fit for very structured businesses (like large corporations, for example).

As we have emphasized throughout this book (as well as the first installment of the series), we thoroughly believe in the fact that every business should find their own agile path. For some, the "traditional" Scrum or Kanban may be enough. For others, however, a more complex combination might be needed. Experiment and see what work for you!

Chapter 6: Dynamic System Development Method (DSDM)

At this point in the book, you have learned about the two most popular agile project management methods (Scrum and Kanban), the second most popular one (XP), as well as a couple of methods that seem to have been forgotten but which can prove their value, especially in a hybrid context (Crystal and FDD).

For the last chapter of this book, we have decided to dive a little deeper into a method that seems to be very complex (and thus, quite scary especially for beginners in the art of agile project management): Dynamic System Development Method (DSDM).

If in the case of FDD the name is quite suggestive of the nature of what the methodology employs, DSDM may appear to be a total enigma when you first look at it.

We understand why it might sound downright confusing and why you might not be that open to learning about it if you are at a beginner or even intermediate level of agile project management. However, we must mention that DSDM is far easier than it sounds.

Like all the methodologies we have approached in this book, DSDM too abides by the general Agile Principles. As a result, it too can be defined as an agile, iterative, and incremental project management methodology. Like many of the other methods we have described here as well, the Dynamic System Delivery Method started off as a framework pertaining to the world of software programming. In time, however, some of its concepts and principles have transferred to other industries as well.

The DSDM approach has been largely developed upon the foundation laid by RAD (Rapid Application Development), a method that lies at the confluence between agile, adaptive, spiral, and unified project management. In essence, RAD is another way of approaching agile project management, but one that comes with its own set of principles and specificities that make it look like it pertains to everything nontraditional in the field of project management in general.

The delimitation between RAD and agile are very hard to make, and, as such, most of the theoreticians are happy to include RAD (and its offspring, including DSDM) under the wide umbrella term known as agile project management. It's not wrong to do this, as the principles are congruent, and there is a common antitraditionalist factor to the entire point of view in both the case of agile project management and Rapid Application Development.

In many ways, RAD represents just a slice of agile when you compare the core of the methodology. It focuses a lot on the speed of production and on delivering working software, and it doesn't focus on procedures and documentation.

What RAD lacks as compared to the other agile methods is the focus on the human element of a development project, as well as the mindset behind it. In this respect, RAD and FDD are quite similar, but it is important to note that DSDM has come as a solution to this because it includes the human element in the entire process and places it at the center of its main principles.

Moving to the actual topic of this chapter, Dynamic System Development is a methodology that sets out the time, the costs, and the quality of the final delivery from the onset of the project. In order to make sure it sticks to the plan, this method splits tasks into four main categories according to their priorities:

- musts
- shoulds
- coulds
- won'ts

One feature of DSDM you might find interesting is that its handbook is available for free, online. Furthermore, multiple resources and templates are also available for download, for those of you whose curiosity might be stirred after reading about the basics of DSDM.

Before we dive into the specifics of Dynamic System Development, we would like to take a moment to analyze its eight main principles: the guiding light of the entire methodology and the core ideas according to which everything DSDM happens.

In short, the DSDM principles you should be familiar with include the following:

- Business needs are very important. It is crucial to understand the specific business need behind the project to be able to deliver what the customer needs.
- The delivery must be made on time, as this helps keep the project on its right course and allows you to keep the customer happy.

- Collaboration is a core value of the DSDM approach, same as in the case of all the other agile project management methodologies out there. You should be able to properly collaborate as a team, as well as with external and internal stakeholders (including the customer).
- The quality of the product should never be compromised. At the end of the project, this is what brings business value both on your end and on the end of your customer, so it is essential to make sure the product abides to the highest standards of quality possible.
- You should build incrementally, but it should all start with a proper, solid foundation. Without that, the "house" will crumble, so take your time in ensuring that the foundation is solid enough for you to start building on.
- You should deliver iteratively, same as with all the other agile project management methods. This allows constant feedback to be properly incorporated into the development process, so that the final product is suited to the client's needs in all respects.
- Communicate at all times. You should be able to communicate clearly and continuously, both internally, as a team, and externally, with your customers and/or other external stakeholders.
- Demonstrate that you have control over the entire process. As a DSDM adopter and as an agile practitioner, you should be on top of the changes that come along the way, and you shouldn't allow them to spiral out of control.

Furthermore, aside from the core principles behind Dynamic System Development, you should also keep in mind that this method is characterized by a series of specific techniques one should use. They include the following:

- Timeboxing. This concept is quite interesting because it will provide you with a mindset that will help you prioritize tasks. In DSDM, the time needed for delivery and the budget needed for delivery are fixed variables. As such, the only flexible variable you are stuck with are requirements.

  This means you should prioritize the requirements according to the importance they have in the overall functionality of the product. If you run out of time or money, you should leave the least important features out of the iteration and deliver the working product as such. As long as the essential 20% of the requirements are satisfied, you can consider your iteration delivery a successful one.

  Do keep in mind that this does not mean you are allowed to deliver an unfinished product. However, the requirements left out of an iteration should be seen as a way to improve the product throughout the future development process.

- MoSCoW is an acronym for what was already mentioned in the beginning of this chapter: work items and/or requirements can be prioritized according to whether or not they *must* be done, *should* be done, *could* be done, or *will not* have to be done.
- Prototypes are an important concept in DSDM, as this method advocates for the creation of system prototypes early in the development process. This will allow you and your team to discover any kind of shortcomings your system may have and

do it early in the development process, so that amends can be applied as soon as possible.

- Testing is another crucial concept in DSDM (one that is common to most of the agile project management methodologies, actually). Continuous testing and feedback allow for the creation of a better product in the end.

- Regular workshops that bring together the different stakeholders of a project are also encouraged. These workshops allow the stakeholders to discuss the requirements, the functionalities, and the specifics of the project and ensure that everything is crystal clear for everyone involved.

- Modeling is a concept that is somewhat common to FDD but with a lack of emphasis on feature modeling and more of a focus on visualizing business domains so that you can improve your understanding of the product and its requirements.

- Configuration management is a practice that helps managers handle multiple deliveries at the same time, at the end of each timebox.

As you have probably noticed thus far, DSDM is a pretty prescriptive approach in the world of agile project management (at least as compared to other methods in the same spectrum). Thus, it makes sense that it comes with a set of predefined team roles as well:

- The Executive Sponsor (also known as the "Project Champion") is on the side of the user organization and has the power to divert funds towards the development of the project.

- The Visionary is the person who initializes the project and draws the basic requirements earlier on in the process.

- The Ambassador User is the person who connects the user community and the project and makes sure that essential information reaches the development team.
- The Advisor User is a user that is actively involved in providing feedback and ideas with relation to the product being developed.
- The Project Manager is either a user or a member of the production/IT Team who is assigned to the management of the project.
- The Technical Coordinator is a person responsible to create and design the system architecture and ensure the technical quality of the product.
- The Team Leader is a person whose main purpose is that of leading their team and making sure that they work effectively and efficiently.
- The Solution Developer is a person charged with the interpretation of the system requirements and their modeling.
- The Solution Tester is the person responsible for checking the correctness of the code, as well as raising the issue when defects are spotted.
- The Scribe is responsible for the collection and recording of the requirements, agreements, and decisions of every workshop. In general, another member of the team will be temporarily assigned to this task.
- The Facilitator is the MC of the workshop—the person who makes it happen and ensures that discussions stay on track.

In addition to these roles, some DSDM projects will also include specialist roles, such as that of the Business Architect or System

Integrator. These roles are not prescribed but recommended under certain circumstances, depending on the complexity of the product and on the complexity of the project in general.

It is also quite important to mention that the DSDM method also ascribes certain factors to the potentiality of success. In short, there are four main points that make a project successful:

- Senior management must accept and support the implementation of DSDM as a project management method. This helps ensure everyone is motivated throughout the entire duration of the project.
- End-user involvement is crucial as well, and for this to happen, you must make sure that higher management is willing to get them involved as well.
- The team must include skillful members. But maybe even more importantly, the team must be constantly empowered, not just in the sense of putting them in the right mindset but also in the sense that obstacles should be removed from their way. For instance, the team should not ask for approval for every small change. Instead, a delegated person in the team should be empowered with a certain level of decision making. This way, time and effort will not be wasted by running paperwork through the management chain of the organization.
- Customers and vendors should maintain a supportive relationship based on communication and collaboration.

In very brief terms, this is what DSDM is all about. Combined with the specific practices of this method's lifecycle (which we will discuss further on in each of the following sections), these principles and rules make for what is generally known as the Dynamic System Development Method.

As you have probably noticed, most of these concepts are congruent with agile project management. So, if you are familiar with its general ideas and Principles, you will find that internalizing the innerworks of DSDM is quite easy.

Let's take a closer look at the Dynamic System Development Method lifecycle and everything it entails.

Feasibility and business study

The first stage of the DSDM lifecycle consists of two main phases: the feasibility study and the business study. They have to happen in this specific order because the feasibility study will provide you information for the business study as well.

During the feasibility study stage, you should study how feasible (or how possible) your application idea is. If it is deemed feasible according to specific criteria, you will have to look into the available team and the available budget. At the end of the stage, you will have to prepare a report on how the product meets the feasibility criteria (time, budget, resources, etc.), as well as a prototype (model) of the project.

During the business study stage, the business experts and the technical experts are brought together to discuss the main issues that might arise throughout the development of the product. The problems will be listed and documented for further reference. Together, the business and the technical experts will determine if they have the business and the technical capabilities to handle the project successfully. If the result of the discussion is a "Yes," the meeting (or series of meetings) will be concluded by a list of requirement priorities, as well as diagrams of the application and product infrastructure.

Functional model/prototype iteration

Once you have all the requirements set in place, the data collected during the predevelopment phase (the first stage of the DSDM lifecycle) is pulled together in a functional prototype of the product. The model will include all the requirements, and it will organize them incrementally.

The prototypes will be further studied and split into smaller substages, including:

- The identification of the functional prototype—the key functionalities you want to include in the prototype.
- The creation and acceptance of the plan and of the schedule.
- Creating the functional prototype: bringing in the programmers and asking them to create a prototype of the product based on what has already been identified and planned in the previous stages. It is important to include a testing phase here as well. Just because this is a prototype, it doesn't mean that it shouldn't be perfectly functional.

- Reviewing the prototype. Once the prototype of the product is ready, it is time to place it in the hands of the end users and ask them to test it. Their feedback and comments will be taken into consideration for future iterations, as the product prototype will continue to be grown and improved throughout the entire duration of the project.

Design and build iteration

The third phase of the DSDM lifecycle is all about building on the prototype and ensuring that it is a continuously improved product. From the prototype, you and your team will move on to developing specific functionalities (also known as individual units) and integrating them into the system.

It is worth mentioning that in Dynamic System Development, there is no clear distinction made between design and build; both of them are handled during this stage of the DSDM lifecycle.

Furthermore, it is important to mention that this stage is also split into four smaller substages, as follows:

- The identification of the design prototype, which includes the requirements that have been decided upon in the prototype/model and then prioritizing them.
- The acceptance of the plan and scheduling. Once the requirements have been planned and scheduled, they need to be agreed upon with the team.

- The creation of the design prototype. Same as in the case of the functional prototype, this stage will deal with the actual development and testing of the design prototype.
- Reviewing the design prototype. At this stage, you will allow the design prototype to be tested and ensure that it is correct and congruent to what was initially planned. Any changes will be implemented in future project iterations until the product is ready for a full release.

Implementation

If the previous stages of the DSDM lifecycle were more about planning and developing small pieces of the product for the purpose of observation, this stage is all about actual implementation.

This is all about watching the live action unfold: putting the product in the hands of the end users and allowing them to fully test it and see if all the business requirements are met.

The Implementation phase is (perhaps unsurprisingly) split into four substages as well:

- Getting the user's approval and offering them guidelines on what should be tested more specifically
- Training users
- Implementing the feedback, you receive from the users

- Reviewing the business needs and whether or not the product meets them. If any new valuable features are identified, they will be further implemented in the product.

Like all agile project management methodologies, the Dynamic System Development Method is based on multiple iterations, so the steps will be repeated for as long as necessary to ensure that the product satisfies all needs and that it is fully ready to hit the market.

As we were saying at the beginning of this chapter, DSDM might sound overly complex at first, but once you nail down its basics, you will understand that this specific method is, just like FDD, a very suitable one for organizations that are more focused on structured planning and documentation. For you as the DSDM project manager, the emphasis will lie on delivering working products. For your organization, however, DSDM and FDD can provide you with plenty of structured information and reporting to keep higher management sound asleep at night.

As you have seen, all agile project management methodologies are more or less variations on the same topic. All of them abide by the Agile Principles. And all of them aim to deliver quality, timely, and budget-friendly products—in software development and every other industry that has embraced agile at the core of their functionality.

Conclusion

In 1971, Ray Tomlinson sent an Earth-shattering message: *something like QUERTYUIOP* (Computing History, 2019).

OK, the message itself was pretty far from life changing. It was quite nonsensical, in fact.

What that message was doing, however, was about to shake the world. Nearly five decades later, we wish that message would have been something more epic, like *a small message for man, a big message for mankind*. But it wasn't. It was just like all those random tests people run when they implement a new system. They make no sense, they come out of the blue, and some of them get to write history in some of the most ridiculous ways possible.

Tomlinson's message was exactly that: a ridiculous experiment that entered history. The message behind the letters written on the screen doesn't even matter today. What matters is that this was the first email ever sent, and it set in motion a series of changes that affect us to the date.

Can you imagine your coffee break without a scroll on Facebook? Can you imagine arguing with your friends on what year the first Terminator came out in and not being able to get on Google and find they were right? Can you imagine not posting photos of your wedding day for all those envious high school classmates to see?

Probably not.

The internet has grown to be such a massive part of our lives that it is impossible to imagine life without it. At work and outside of work, we are constantly relying on the internet to provide us with the connection we need. We rely on it to connect us to information, entertainment, and downright bitterness over the decades that have passed since our high school graduation.

What Tomlinson did mark the beginning of a whole new era in the evolution of the internet. Sent via ARPANET (a primordial ancestor of modern-day internet), his totally nonsensical message was, in fact, the beginning of online communication as we now know it.

It took another couple of decades until the internet got in every home, but beginnings are always timid like that. Once set in motion, the machinery powering the bytes behind the screen blew up to an entire phenomenon, one we now know as the (in)famous dotcom bubble.

On this background, software development companies were facing one of the most gruesome and most irritating issues in the world of tech: lag. They were lagging so much behind schedule that some of them delayed product releases by *years*.

The real issue was not that there wasn't enough talent or that these companies were lazy in any way. The real issue was connected to the fact that all these projects were very volatile in nature. Requirements were changing all the time, new technologies were emerging all the time, and the very nature of these programmers' jobs was changing from one day to another.

And yet, they were all stuck with strict management methods that had absolutely no idea how to deal with the massive amount of uncertainty dominating the software development world.

Everyone was in dire need of change. They needed a framework that would be able to actually help them respond to all these shifting situations.

One by one, lightweight agile methods started to pop here and there. Crystal, Scrum, and XP were among the first ones. And before anyone realized what was going on, the Japanese came from the far East to put on the table a solution they had been using for several decades by then: Kanban.

From a world of strict rigor and nearly obsessive planning, the world of software programming was blown up by the chaos of all these innovative, albeit unregulated new methods.

And then, in 2001, a bunch of smart guys met up for a ski trip, sat down, and laid the foundation of the Agile Manifesto: a framework that brought together all the lightweight methodologies and unified them under a common set of Principles.

Magic history is rarely planned. The first email sure wasn't, and it is more than likely that the creation of the Agile Manifesto wasn't a thoroughly planned moment either. There were no confetti popping in the background of these moments. No fireworks. No special concerts. Nothing to say that those were special days in the evolution of mankind.

And yet, these moments happen. They happened behind locked doors and as a result of hard work and determination and frustration. They happened because it was the only natural thing to do. They happened. And they changed the world.

By the end of this book, we truly hope you have come to grasp the full agile project management grandeur in all its wonderful formats and colors. We hope you have come to understand that good planning is not about knowing what will happen, but about knowing how to respond to what you don't know yet.

The best moments in mankind's history are moments of shifting change, moments that happen silently, without much buzz around them. What would we be as a race if we didn't know how to adapt to these changes?

You see, agile is the most natural thing in the world. Mankind has been inadvertently doing it ever since we started using tools and making fire. And then, we somehow stopped doing it naturally and started closing ourselves and our resilience in the face of change behind strict rigors and obsessive-compulsive planning.

Agile project management is one of the most natural paths you can take, especially in the ever-changing environment of today's world. It's not even a matter of adapting to the slow changes that happen behind locked doors anymore: it is a matter of adapting to massive changes in everything around us.

From climate to politics to society, our world is drifting apart from its traditional views and opening the gates to a whole new future—one only science fiction writers could have ever imagined.

The wonderful thing about everything we have presented in this book (as well as the first installment of our Agile Project Management series) is that *agile* has been the motor of change for quite some time, and it will continue to drive change as more and more companies from wildly varied industries start embracing it.

The more flexible you become, the easier it is for you to truly embrace what comes ahead, and agile project managers know this better than anyone. Being inflexible, on the other hand, will only lead to broken bones and painful failure.

We hope this book was able to provide you with a full, comprehensive overview of some of the most popular agile project management methodologies in the world. We do not claim we have said it *all*. It would be completely absurd for us to say that. Instead, we claim that our book is one of the best pieces you will ever read on the topic of how the main agile methodologies connect into their agile mothership, and how they connect to each other.

As we said in the introduction, we cannot tell you the *right* answer. Agile may or may not be for you because yes, it is a perfectly valid point that agile is not for every single company out there.

What we do hope is, however, that you will give agile a chance. Even in its strictest forms and even in hybrid formats, agile can still be the internal change you need to be able to withstand the external shifts in the world.

Our journey through Scrum has shown you that there is a very good reason this method is very popular but that you should look beyond the apparent playfulness of the practices employed in this approach as well.

Our journey through Kanban has shown you that something can be old and new at the same time, and that agile does not always originate in the realm of software programming.

Our journey through Extreme Programming has shown you that sometimes, popular methods fall back into near oblivion, but they continue to survive through what they do best: the specific practices that have been borrowed by other project management frameworks.

Our journey through the Crystal Method has shown you that some of the largest companies in the world can become agile too, despite all the apparent challenges they had to face in the process.

Our journey through Feature-Driven Development has shown you that it doesn't matter how complex an agile method may look from afar: when you strip down, it's all about the flexibility.

Finally, our last journey through the Dynamic System Development Method has shown you that yes, agile can be documented and reported as well. So yes, it can definitely work in organizations that treasure these features too.

Overall, we hope that this entire journey through the agile project management methods of the current landscape has shown you that there is no right or wrong solution, and that most of the times, the right path lies in knowing which practices are specifically good for you, your team, and your business.

Agile project management can be genuinely fascinating when you know how to look at it. It might sound like the dullest topic in the world when you work outside of it, and it may seem like it's an endless discussion on which spreadsheet formula works best for reminding people that they have a deadline to attend.

But when you look past all these misconceptions, you will discover that agile project management is all about the marvelous change we have all witnessed in the past few decades.

We truly believe we are not exaggerating when we say that agile has been the engine of change. Behind the great ideas of Steve Jobs and Mark Zuckerberg, behind the recent launch of a car in space by Elon Musk, and behind the most proficient businesses in today's world, agile works its way in and out of trouble. *In,* because it doesn't fear change and issues in the process. And *out* because it is the very core of the problem-solving mindset, we all need to adopt.

At this point, it doesn't even matter if you are a project manager, an entrepreneur, or simply someone who wants to achieve the best version of themselves.

All that matters are that you can take the agile methods we have described here, nitpick them, test them out, and see what works best for you. We guarantee that you *will* find the right formula to success, regardless of whether that is all about delivering a working software program or growing a pair of triceps by next summer.

Agile can change you inside and out because it shifts your entire perspective on life and how you should react to the changes around you. And for this reason, we cannot be anything but proud to say that yes, *agile rules the world.*

We wish you luck in your future endeavors, and we truly hope this book has been of help for you, for your team, and for your business. It won't be an easy journey ahead, no matter what you decide to do next, but we truly hope that this book has taught you that resilience, hard work, and pure determination can make even the most agile-adverse situations turn around and smile back.

At the end of the day, agile is all about continuously iterating your product's releases until you get it right and until you find the (perhaps not so secret) formula to success. And your very journey into agile should be a metaunderstanding of everything you have learned about this project management mindset and everything it brings along!

Stay strong, stay smart, stay curious, stay agile, and the Universe will smile down upon you!

Printed in Great Britain
by Amazon

34988650R00151